MARINE
CONSERVATION
IN FIJI

MARINE CONSERVATION IN FIJI

CORNELIUS TYSON

iUniverse

MARINE CONSERVATION IN FIJI

iUniverse books may be ordered through booksellers or by contacting:

iUniverse
1663 Liberty Drive
Bloomington, IN 47403
www.iuniverse.com
1-800-Authors (1-800-288-4677)

ISBN: 978-1-5320-2150-3 (sc)
ISBN: 978-1-5320-2151-0 (e)

Print information available on the last page.

iUniverse rev. date: 04/20/2017

DEDICATION

I would like to dedicate this book to my Biology Professor Dr. Jarrett at Central Connecticut State University for his advice and guidance on the project that led to my writing this book, Global Vision International for the opportunity to work with them in Fiji, and especially my parents who provided me with love, support and just enough pushing to set me off on the right path to pursue my dream career.

CONTENTS

DESCRIPTION OF PROJECT

In 2015 and 2016, the author participated in a community-based marine conservation project in Fiji sponsored by Global Vision International (GVI). He collected initial monitoring data on marine ecosystem species around the island of Caqalai. This data was used to assess existing conditions of the ecosystem and to help create conservation policies and educational strategies for the islanders of Caqalai. In addition, he spent a significant amount of time working with an organization on the island of Viti Levu that specializes in shark conservation.

This book consists of three sections: 1) research methods for observing marine species, 2) marine species of Fiji, and 3) community based marine conservation strategies. Each section includes a review of the literature on the subject and a summary of his experiences while in Fiji. The information presented includes explanations as to why each animal, from the smallest coral polyp to the largest bull shark, play an important role in maintaining the ecological balance of reef ecosystem. This book is intended to be used as a field guide for those interested in marine conservation in the South Pacific.

THE SITUATION IN FIJI

Much like the majority of countries in the South Pacific, Fiji is home to a vast diversity of coral, fish, invertebrate and other marine species. According to the Global Vision International (2015), "there are hundreds of species of hard and soft coral, sea fans and sponges, a recorded 1,200 species of fish, nearly 500 species of mollusk including over 250 nudibranchs and over 100 bivalves in Fijian waters. Four species of marine turtles visit the waters of Fiji. Humpback whales pass through during their austral winter migration, and in some places there are resident Spinner dolphins." Fijians rely on these organisms for food and as tourism attractions, they provide the foundation for much financial income, and are the basis of much of their spiritual folklore.

Unfortunately, due to overfishing, climate change and nutrient pollution many of these species and their habitats are in danger. These problems are made worse by the growing demand put on marine resources. The demand for more food rises as the population of this island nation increases. From 1996 to 2016 the population of Fiji has increased from 775,077 to 897,537. With an increase of over 100,000 people in 20 years, the demand for food has grown quickly. Because of such sharp and rapid growth, the sustainability of these marine resources is under heavy pressure. Pressure is also coming from outside sources, mainly demand for food products for the Asian delicacy market. Shark fin soup and various sea cucumber dishes like Bêche de Mer are viewed as a delicacy in many parts of Asia. The rise in demand for these products is causing wide spread poaching and is decimating shark and sea cucumber populations. In order for Fiji to build a case for better managing their marine resources, they must carefully monitor the conditions of these resources. Monitoring data is used to develop policies that aim to protect marine species under threat and their habitats. In addition, this information is used to create community programs to help educate people about new fishing and development strategies that better protect the marine environment.

RESEARCH METHODS FOR OBSERVING MARINE SPECIES

LITERATURE REVIEW

There are many research methods used to survey the marine ecosystem. In Fiji there are three main methods that are used by the Department of Fisheries and the Fiji Locally Managed Marine Area Network (FLMMA) (Global Vision International, 2015). The three research methods are Line Intercept Transect (LIT), Invertebrate Belt Transect (IBT) and Underwater Visual Census (UVC). There are also Opportunistic Surveys that are used to look at species on the International Union for Conserving Nature (IUCN) list of endangered species. The data that is collected by these survey methods are then entered into color-coded GIS maps and used in scientific reports and in environmental presentations with local communities and with the Fijian government.

THE LINE INTERCEPT TRANSECT (LIT)

This method is mainly used to record changes in size or population of plants, coral or other benthic organisms. It is known as the "standard method" of surveying coral reef colony size and health by the Global Coral Reef Monitoring Network (GCRMN) (Hill, 2004). This method is generally done by creating and running various line transects at a set distance and noting the diversity of species, quantity of each species, size and health of the plant, coral or benthic organism that is crossed by each line (Caratti, 2006). The recommended length of each line transect and the length of transect used in Fiji is 20 meters (Caratti, 2006; Global Vision International, 2015). For coral, data is recorded in ratios of heads of coral per the length of the transect (Mumby, 2004).

THE BELT TRANSECT

This transect method is generally run by one person at the same time as a Line Intercept Transect or Point Intercept Transect is run by their partner. This method is used to observe targeted fish and invertebrate species abundances.

In Fiji, the use of the Belt Transect method is mainly used to assess the population count of invertebrates. The main focus is on assessing populations of Crown of Thorns Sea Star (*Acanthaster planci*) and several species of Sea Urchin and Gastropods like the Coral Snail (*Coralliophila neritoidea*) and Drupella (*Drupella cornis*). All of these species are coral eaters and when populations have gotten too high they are known to cause serious harm to live coral populations. Other invertebrates that are assessed may include: Trochus (*Trochus niloticus*), Triton's Trumpet (*Charonia tritonis*), Tiger Cowrie (*Cypraea tigris*), Giant Clam (*Tridacna gigas*), Sea Cucumbers, Painted Crayfish (*Panulirus versicolor*), and the Pronghorned Spiny Lobster (*Panulirus penicillatus*). These species are being harvested for their shells that are made into jewelry, sold as souvenirs, or sold as high priced food delicacies in Asia or Fijian Resorts (Global Vision International, 2015).

The diver that runs the Invertebrate Belt Transect is the dive buddy of the diver running the Line Intercept Transect as these can be run at the same time with the same set up. The difference is while the Line Intercept Transect diver swims straight along the 20 meter transect line recording the benthic organisms it crosses, the Invertebrate Belt Transect diver does a zigzag pattern along the transect line. This zigzag motion strays 2.5 meters to the left and the right of the transect line. This allows the diver to observe invertebrates that are not in clear view from right above the transect line (Global Vision International, 2015).

THE UNDERWATER VISUAL CENSUS (UVC)

The Underwater Visual Census is another method used to observe and record fish populations. There are three methods of implementing an Underwater Visual Census each with their own sets of pros and cons. There are the Capture Method, Non-Capture Method, and the Mixed Method.

Capture Methods are methods that observe the numbers, health and size of fish species that are caught by traps, nets and fishing lines in a specific area. Data are recorded in units of catch per unit of effort (CPUE). Data collected using this method can be used to determine the population density of an area. There are pros and cons to using this method of data collection. Because there is no real training required for recording the number of fish caught, local fishermen can be hired to do the data collection instead of trained researchers. This has the advantage of making the price of research significantly cheaper. Yet, there are often many variables that cannot be controlled that will bias the outcome of the research. Because you must first catch the species being collected, the types of baits, fishing equipment and the "capturability" of each species become significant variables that can affect the accuracy of the data (Labrosse, 2002).

One other type of Capture Method that has a more definitive outcome is the use of explosives and poisons (i.e., dynamite and Rotenone). While this method is very accurate in determining population density, it is obviously an extremely destructive method of data collection.

The Non-capture or 'fishery independent' method is done in the water using snorkel or scuba equipment with a transect set up on the reef. The standard is a 50 meter transect with a five meter^3 diameter (Global Vision International, 2015). Snorkelers or divers swim along this transect counting the number of fish and the number of fish species, and record the size and whether or not they are juveniles or adults. Because of this, this method requires accurate knowledge of fish species, their life cycle, and diving experience. The level of training needed for this method is much higher than the Capture Method. Another constraint that can hinder this method is water visibility. While this method can be extremely accurate with good visibility, poor visibility can make this method inaccurate or even impossible to conduct (Labrosse, 2002). The Underwater Visual Fish Census Survey (2002) states that this method is best carried out in clear, calm and shallow water for the most accurate representation of population density in an area.

The third type of Underwater Visual Census method is the Mixed Method. Mixed Methods are a combination of the previous two methods. This method is very useful for monitoring fish migration and behavior patterns.

There are two main types of the Mixed Method. The first is to capture a fish, record its species, gender, and size, tag it with a GPS tracker, and release it so that fish's movement can be monitored. The second type is similar, but the fish is not captured, this is normally done with larger organisms like whales. Instead of capturing the organism, a diver would approach the animal in the water and attach a GPS tracker to it. This is typically done with a spear gun or sling. The GPS data is used to determine migration patterns, primary hunting grounds and breeding locations of the tagged organism. This data can then be studied to see how coastal developments, climate changes, over fishing and other factors are affecting these organisms.

The following is a table that summarizes the effectiveness of the various research methods that have been discussed.

Sampling Techniques	Comprehensiveness	Accuracy	Coverage	Bias linked to life cycle	Staff Training	Costs
Capture	Low*	Low to moderate	High	Yes	Low	Low
Non-Capture	High	High	Low	No	High	Moderate
Mixed	Low	Moderate	Moderate	Yes	High	High

*Except for explosives and poisons
**Table found in Labrosse (2002)

EXPERIENCE IN FIJI

GVI's research methods utilize teams of five volunteers. Each volunteer has a specific job. They are thoroughly trained in and out of the classroom by GVI to be able to accurately identify fish, coral and invertebrate species in their various stages of development. They are tested on paper and in the water by instructors to ensure they can collect accurate data.

These five man teams have a standardized template they use to conduct these surveys. Each survey site is marked with GPS and at each of the sites there are two depth ranges in which surveys are carried out, 3-5 meter and 8-10 meter. At each depth, 5x50 meter transects are completed using the UVC method. In addition, 5x20m transects are completed at each depth using the LIT/IBT methods. The transect line is laid across the reef surface at a constant depth, usually perpendicular to the reef slope. Each team of researchers consists of the following:

LEAD DIVER/PHYSICAL SURVEYOR

This is the leader of the survey team. They are in charge of setting up each transect (50 meters long with a 5 meter^3 diameter) at each of two specified depths on the reef (3-5 meters and 8-10 meters). This person's role is reserved for the most experienced diver of the team as their buoyancy control must be good enough to be able to lay down the transect line (a 50 meter tape measure) while avoiding damage to corals. On a wall/drop off survey they would place the transect line along the wall between 3-5 meters or 8-10 meters rather than over the bottom. During surge this can be difficult as the tape measure can become slack, move up and down with the surge and make it extremely difficult for the LIT surveyor. In order to avoid this, it is important to keep the tape measure tight while it is being laid down and hooking it in crevices or carefully around coral. It is important to note that when laying the transect tape it is preferable to continue in as straight a line as possible. However, in a situation where the substrate declines or rises beyond the targeted depth, it is important to follow the constant substrate. Another important note is that if there are two or more surveys going on during the same dive, the randomness of the survey decreases. If both survey teams are diving the same area, they cannot go in the same direction as this could alter the number of fish or invertebrates in their transect areas. Therefore, having multiple survey teams in one survey area at the same time will decrease the randomness of each survey. Each survey should remain as random as possible in order to decrease the amount of bias in the data.

The Lead Diver/Physical Surveyor is also required to record the surface temperature, surface salinity, bottom temperature, bottom salinity, time, visibility, start depth and end depth of the transect on a slate. This is done to study any relationships between changes in environmental variables and the fish, coral and invertebrates being observed. Once they have completed their primary duties, they act as back-up and help their dive buddy, the Underwater Visual Census Surveyor, record their observations. They do this by swimming behind them and making sure they stay at the correct depth and direction of the transect. In surveys that take an extended length of time (i.e., surveys along walls/ drop offs), once the Physical Surveyor completes their tasks they can also assist the Line Intercept Surveyor. They can do this if they have received training as a Line Intercept Surveyor. They can start at the opposite end and move towards the starting point.

UNDERWATER VISUAL CENSUS SURVEYOR (UVC)

This volunteer is paired up as the dive buddy with the Physical Surveyor. The Underwater Visual Census Surveyor swims with the Physical Surveyor behind them in the direction the Physical Surveyor indicates. As the physical surveyor lays down the transect line, the UVC surveyor observes all fish in the immediate three meters to the left and three meters to the right of them. They record the species, development stage and quantity of each fish seen on their slate. They also categorize each fish observed by size. They do this by putting them in groups of increments of five cm ranging from 0-5 cm to 40cm+. It is important to note that the increments start at the next centimeter up; i.e., 6-10 and 11-15 rather than 5-10 and 10-15, as this decreases the personal influence that each surveyor has while surveying. For example, two surveyors could place a 15 cm fish in either a 10-15 or 15-20 category; however if the categories were 11-15 or 16-20, the decision would be more accurate and less arbitrary. This is repeated at each transect at each specified depth. This data is taken from each of the 50 meter transect areas.

A typical UVC surveyor's slate would look like this:

	0-5	6-10	11-15	16-20	21-25	26-30	31-35	36-40	40+
Parrotfish		2**	3	8	11	16		2	
BMB*		2	4	12					
Humphead Wrasse									1

Fusilier			350***						
LBT*	2	1	8	1					
Red Snapper					5		15	25	

*To save time fish species names are often abbreviated during the survey and extended to full during post-survey data entry (e.g., Bridled Monocle Bream is shortened to BMB or Lined Bristle Tooth is shortened to LBT).

**Replace numbers with tally marks on surveys.

***When large schools of schooling fish (e.g., Fusiliers) are observed it is impossible to get an exact number, but estimating by counting by groups of 5s or 10s is acceptable.

LINE INTERCEPT SURVEYOR (LIT)

The Line Intercept Surveyor is charged with the task of observing and recording the species, abundance, health and estimating the size of each coral, other benthic species or substrate (i.e. sand, rubble, rock and dead or bleached coral) that are within the transect area. While the Underwater Visual Census Surveyor does a 50 meter long transect, the Line Intercept Surveyor only retrieves data from a 20 meter transect line. They do this by swimming behind the Physical Surveyor and the Underwater Visual Census Surveyor along the transect line, looking down and recording on a slate the species, the substrate and the size of everything that falls beneath the transect line. To determine the size of what lies beneath the transect line the LIT surveyor looks at the measurement on the tape measure and records where the benthic life form or substrate starts and ends. The measurement of where it ends would be the start of the next benthic life form or substrate and the same would be recorded for everything under the transect line until the 20 meters is reached. This action is repeated at each transect at each specified depth. The data is then used to create an estimate of coral bottom coverage. When surveying and entering the surveyed data it would be impractical to write out the benthic life form or substrates' full name. A series of codes have been created to streamline the surveying and data entry process. These codes are found in the table below.

Benthic Lifeforms

Coral	Lifeform	Code	Description
Acropora	Branching	ACB	at least two degrees of branching (secondary branching)
	Digitate	ACD	single branches with some second degree
	Tabular	ACT	horizontal flattened plates
Non-Acropora	Branching	CB	at least two degree branching
	Encrusting	CE	encrusts onto substrate
	Foliose	CF	coral attached at one or more points, leaf-like
	Massive	CM	solid boulder or mound
	Submassive	CS	tends to form irregular shapes, small columns, knobs or wedges
	Mushroom	CMR	solitary, free-living
	Millepora	CME	fire coral
	Tubipora	CTU	organ-pipe coral
Dead Coral		DC	recently dead, white to dirty white
Dead Coral with Algae		DCA	this coral is still standing, structure can be seen but covered with turf algae
Other Fauna			
Soft Coral		SC	soft bodied corals
Sponges		SP	

Zoanthids		ZO	
Others		OT	tunicates, anemones, gorgonians, sea whips, giant clams
Macro Algae		MA	
Turf Algae		TA	
Crustose Coraline Algae		CCA	
Sand		S	
Rubble		R	
Silt		SI	
Water		WA	defined as 50 cm above the nearest object
Rock		RCK	

* Table found in Global Vision International, 2015

A typical LIT surveyor slate after a survey would look like this:

100*	ACB	240	OT	816	CCA	1545	CE
135*	CM	258	S	834	CE	1599	S
202	ACB	604	MA	1032	W	1720	MA
212	SP	632	S	1342	ACT	1845	S
218	CE	660	MA	1421	CCA	1900	MA
225	SP	809	OT	1434	MA	1955	DCA
230	CE	813	TA	1521	OT	1995	S
232	SP	815	OT	1530	CCA	2100	MA

*The LIT survey starts at 100 cm and ends at 2100 cm because the first meter is used to anchor the tape measure to the bottom using a weight or tying it around a rock or piece of dead coral.

**Each number represents the start of the corresponding object surveyed and the end of the previous. Therefore, 100 ACB that ends at 135 CM represents 35 cm of Acropora Branching and the start of the measurement of Coral Massive.

INVERTEBRATE BELT TRANSECT SURVEYOR (IBT)

The Invertebrate Belt Transect Surveyor is paired up as a dive buddy with the Line Intercept Surveyor. They are in charge of observing and recording the species and abundance of targeted invertebrates. The targeted invertebrates include coral predators like the Crown of Thorns Sea Stars and Drupella Snails as well as invertebrates that are being over harvested like the various species of sea cucumber and giant clam. Like the Line Intercept Surveyor, the Invertebrate Belt Transect Surveyor only retrieves data from a 20 meter transect line. They do this by swimming in a zigzag movement to the left and right of the transect line, while still staying inside the transect area. On a wall/drop-off survey instead of going left to right by 1.5 meters, the IBT surveyor would zigzag up and down by 1.5 meters. The zigzag movements allow the surveyor to observe organisms that could not be seen from a straight down view, they are encouraged to look into crevices and along the base of certain corals to find hard to see invertebrates. For example, Drupella Snails are often found at the base of or in clusters along the branches of Acropora corals (Cumming, 1999). This action is repeated at each transect at each specified depth. It is not uncommon for the IBT surveyor to find little or nothing on a survey depending on the benthic composition, however a wall/drop off survey often produces an excessive amount of small black urchins and in general a higher number of invertebrates.

A typical IBT surveyor slate after a survey would look like this:

Flower Fish	3*
Crown of Thorns	1
Mega Bum	1
Drupella	6
Small Black Urchin	138

*Replace numbers with tally marks on survey

BOAT MARSHALL

The Boat Marshall is the only team member not in the water. Their main job is to make sure the divers are safe while underwater. This includes taking roll-call, helping divers put on gear, briefing divers before they enter the water, watching for bubbles and Signal Marker Buoys (SMBs) to keep an idea of the divers' location, helping divers back in the boat after the dive and recording their dive details before and after the dive. Dive details include: what BCD and regulator they are using (to know what's wrong with them if repairs are needed), the amount of air they are entering and exiting the water with, their maximum depth and time of their dive. In a case of decompression sickness, all dive details are required by emergency services for selecting the best plan of action. The boat marshal enters the dive in a master dive log kept by the GVI staff. On a survey dive, the Boat Marshall is the 5th person on the survey team. They are also in charge of observing turbidity, surface impacts, cloud cover and boat traffic in the general area as well as collecting any opportunistic data.

A Boat Marshall plays a key role in opportunistic surveys. After recording the dive details at the end of the dive, they also record any opportunistic data that might have been collected. This opportunistic data is also entered in the master dive log. Opportunistic data is defined as any sightings of any of the specific species that GVI is particularly concerned with. More details on opportunistic surveys and the species GVI is concerned with can be found below.

OPPORTUNISTIC SURVEYS

As the name implies, Opportunistic Surveys collect data when the opportunity to site a specific species presents itself. This opportunity could be on any dive, regardless if it is a survey dive or a recreational dive. This makes every dive GVI does a potential opportunistic survey.

Most of the species that GVI commonly collects opportunistic data on are species that are found on the IUCN watch list. These include Giant Clams, Humphead Wrasse, Bump Head Parrotfish, Squaretail Coral Groupers, Black Tip Reef Sharks, White Tip Reef Sharks, Giant Reef Rays, Blue Spotted Rays, Eagle Rays, Green Sea Turtles and Hawksbill Turtles. If a GVI volunteer or staff member spots any animal that is on the IUCN watch list it is recorded and added to the count. Every month the data is then submitted to IUCN. This supports their efforts to get an accurate estimation of endangered animal populations. GVI also uses opportunistic surveys to take special notes on populations of Crown of Thorn Sea Stars and their predators such as the Humphead Wrasse and Triton's Trumpet. These creatures are key factors in determining reef health.

PROS AND CONS OF GVI'S RESEARCH METHODS

There are a few pros and cons to GVI's five man team research style. For instance, many research organizations require each of their research teams to be able to identify up to 180 to 300 fish species. This requires months of training. This is not possible in the relatively short amount of time each GVI volunteer is in Fiji. GVI minimizes this limitation by focusing on the smaller number of species that the Fijian communities care the most about and the ones that are used as the main indicators of reef health. GVI reduces the volunteer data collector's required knowledge of fish species down from 300 to 57 fish species. GVI also requires that volunteers have the ability to identify 29-targeted invertebrates and 25 benthic life-forms. GVI divides up this responsibility by assigning each team member certain species to be identified. For example, an Underwater Visual Census Surveyor is not going to be required to be able to identify Crown of Thorns Sea Stars if their job is to identify and record fish species. By assigning each volunteer specific species, GVI focuses their training in a way that maximizes the effectiveness of time spent actually collecting data. Each team member specializes and becomes very good at their specific job. However, because each team member is focused on their own job, there is no way to detect if a mistake is made during data collection (except for the underwater visual census surveyor as they have the physical surveyor as back up). If a mistake is made, all of the data collected during that time should be void. This being said, if GVI's training is carefully followed there should little problem with the accuracy of the data collected. The exception to this concerns GVI's opportunistic surveys. The problem is that without tagging there is no way to know if everyone is seeing the same species. It is very likely that people will see the same resident shark, ray or fish at a specific dive site or that everyone will see the same few giant clams. This makes data overly optimistic and somewhat inaccurate.

ANALYSIS OF DATA COLLECTED AT GVI'S CAQALAI ISLAND BASE

After data collection is completed the data is entered, by hand, from slates to printed papers designed for data entry and from there into a Microsoft Excel spreadsheet. Separate spreadsheets are made for fish data, benthic data, invertebrate data and corresponding sorting spreadsheets. These sorting spreadsheets are used to total the abundance of specific species and organize them into groups, e.g., total abundance of food fish, commercial invertebrates or Algae cover. As with any hand-entered data, mistakes are expected due to hand writing issues, typos or missed-entry of data into spreadsheets. However, these mistakes are often relatively insignificant. Large mistakes are very noticeable and can always be traced back to the original data entry paper for correction.

GVI USE OF GEOGRAPHIC INFORMATION SYSTEM (GIS)

A Geographic Information System (GIS) is a method of analyzing and displaying cartographic data (Itami & Raulings, 1993). It is used to create maps that can portray survey data geographically. GIS can layer many different types of data on a single map of one location; e.g., hard coral locations, resident fish populations, topography, water temperature.

For instance, GVI uses GIS to combine satellite imagery and survey data to create easy to understand color-coded maps that help relationships to be detected between several variables. For instance, results of surveys at 3-5 meters and surveys at 8-10 meters can be compared using GIS maps. The color codes on these GIS maps can be used to show population numbers of all fish/invertebrates or benthic species surveyed, the population of a specific species of fish/ invertebrate or benthic species, or just species of importance such as commercial food fish or invertebrates.

This can be shown for every survey site, just sites on a certain part of reef system, or only survey sites that contain a specific species of coral, algae or another species of fish. The possibilities and combinations that can be shown with GIS are endless.

For example, if data was needed to look at the correlation between populations of Humphead Wrasse and the Crown of Thorns Sea Star, but only when there is a presence of Non-Acropora Submassive hard coral, at a depth between 3-5 meters, and only on the southern half of a surveyed reef, this can be done using GIS. This is a very specific, but very realistic example of how GIS is used. GVI mainly uses GIS maps to show population abundances of commercial food fish, coral predators, algae grazing fish, coral cover, algae cover, and fish, invertebrate, and benthic species diversity.

A satellite map of Caqalai Island and the surrounding reef is used as the base for the GIS map. The survey sites are located along this reef and wrap completely around the island. Points on the base map represent the center of each of these survey sites. Each of the survey sites has an area of 250m^2. The shading around each point represents the area of each site. Because the sites are all next to each other, the borders of the shading connect to form a solid ring around all of Caqalai. The colors of the shading are what are used to display the data collected. GIS uses a gradient of shading in different colors and shades of colors to represent the degree of abundance (or lack of) of the species that is being represented. The darkest colors often represent a high abundance. Low abundance is represented by the lightest color. For example, when looking at populations of Humphead Wrasse around Caqalai, a survey site that has the highest population of Humphead Wrasse surveyed would be represented in dark red, while a survey site with lower abundance might be represented in orangish red. A site with a small population of Humphead Wrasse would be represented with yellow and a site that surveyed a population of little to none would be represented with green. This color gradient of red to green is fairly simple when there are only four survey sites. With more sites and more differences in the findings of populations the color gradient becomes more detailed. The greater the number of colors and shades in the gradient, the more detailed the GIS map becomes. A key is included on the GIS map to show the relationship between data being looked at and the color gradient on the map.

MARINE SPECIES OF CAQALAI, FIJI

INTRODUCTION

Fiji has a very biologically diverse marine ecosystem - 1198 different species of fish, 1056 different species of marine invertebrates and about 1000 coral species have been identified (Fiji Department of Environment, 2010). However, this diverse ecosystem is currently in danger from overfishing, climate change, nutrient pollution and poaching.

In order to maintain the health of Fiji's Marine ecosystem, a firm grasp of the dynamics of Fiji's coral, fish and invertebrate species must be understood. One must know identifiable characteristics of each species, their stages of life and what each stage looks like, as well as their normal behavior including breeding patterns, hunting patterns, what depth range they normally stay in and how mobile they are. The project described in this paper focuses on economically and biologically important fish, invertebrates and coral/benthic species that are either, a) threatening the ecosystem, b) being threatened by over fishing, introduction of an invasive predator, or through diseases, or c) are an important indicator species for the health of the ecosystem.

Cornelius Tyson

LITERATURE REVIEW

IDENTIFICATION CHARACTERISTICS

There are certain characteristics of fish species that researchers use to identify those that are targeted when surveying their populations. These are discussed below.

Fins

The fins of a fish are an excellent indicator of what type of fish the observed species is. The shape and size can be used to determine the speed at which the fish swims, whether it swims at a constant speed or at short fast bursts. If the fin is stiff and hard, that indicates it is a pelagic fish, like a Jack or Barracuda that uses "strong propulsion" to swim. On the other hand a soft and flexible fin is an indicator of a slow moving maneuverable fish, often an herbivore or coral grazer, like a Parrotfish.

Dorsal Fins are located at the top, along the spine of the fish. There can be multiple dorsal fins spaced out along the top such as with certain species of Cardinal fish like the Tiger Cardinal. There may be a single fin at one part of the top like with the Yellow Spotted Trevally; or a single fin ranging across the entire fish like with Angel fish. Some fish also have spikes along the dorsal fin that are used for defense. An example of a fish with this characteristic is a Lion Fish, a predator species native to the South Pacific.

The caudal or tail fin is another characteristic that can be used to distinguish one species of fish from another. The tail is where the majority of speed comes from and because of this, the shape of different species' tails are different based on their needs (Snyderman, 2009). A thin crescent moon shaped tail is a characteristic found mainly in fish like

Trevali and Tuna. It allows these fish to swim at high speeds for long distances in open water. This type of tail is typically not found in slow moving, agile, reef fish. A rounded tail is common in reef fish like Groupers and Sweet Lips. This tail allows for quick movements as well as quick acceleration, but they cannot hold these speeds for very long due to the drag created by the shape of the tail. These characteristics are important for fish that live primarily along the reef but sometimes venture out of it. An emarginated tail is similar to a rounded tail in terms of movement and acceleration but has less drag because there is a shallow notch that makes it slightly forked. Fish with this type of tail can hold higher speeds for longer. Snappers are an example of fish with this characteristic. Forked tails may be larger at the top lobe of the tail or symmetrical on both lobes. Tails that are larger at the top lobe tend to force the fish to swim downward. They compensate by extra use of the pectoral fins to lift them up. This characteristic can be noted in many sharks like Reef, Bull and Hammerhead sharks. On the other hand sharks like the Great White and Mako have symmetrical lobes on their tails. This characteristic allows them to be the fastest in the shark family.

A fish can also have a continuous tail that stretches the entire length of the body without breaking. This is found in fish like the Moray Eel and are used for movement through the cracks and crevices of the reef.

The anal fin begins near the anal opening of the fish and can extend towards the tail. Species like the Moray Eel have an anal fin that fuses with the tail fin and the dorsal fin to create one long fin that goes along the majority of the body.

The pectoral fins are located on either side of a fish behind the gill plate cover. These are used mainly for movement by reef fish and are more prominent in fish that do not use their tail fins as their main source of movement. Depending on the species and the speed that the species swims and turns, these fins can be either very prominent or sleeker and less obtrusive. A fast moving fish that ventures out of the safety of the reef like a Trevally or Jack will have a sleeker less prominent set of pectoral fins in order to swim faster. While a slow moving fish that often hovers in and around the safety of coral heads within the reef (e.g., Lionfish) will have wide flowing pectoral fins that allows it to move and hover around a specific area.

Pelvic fins, like pectoral fins, are located on either side of the fish. Pelvic fins are located beneath the pectoral fins at the bottom of the fishes' body. These fins are often used for stabilization. When in a current, a fish uses its pelvic fins to stabilize itself and bring its body back to a steady swimming posture (Standen, 2008). Like pectoral fins, pelvic fins are more prominent in fish that often stay in the safety of the reef. Once again, Lionfish species are a prime example of this as it constantly uses its pelvic fins to stabilize itself while it hovers.

Mouth

The location of the mouth, the direction it points, the extent the fish can open its mouth, and the teeth the fish has, are all indicators of the species of the fish and what it feeds on. For example, a Stingray has its mouth located on the bottom of its body which indicates it primarily eats bottom dwelling creatures like small fish, crabs and mollusks. The more that a fish can open its mouth, the larger the prey that fish can consume. For example a Tiger Shark can open its mouth quite wide. Another feature is the type of teeth a fish has. Sharp pointy teeth, like those on a Shark, Barracuda or Trevally indicate a carnivorous predator. Flat fused beak-like teeth are indicators of coral grazers that use them to bite off and grind pieces of coral (Fox and Oxford Scientific Films, 1982). An example of a fish with these type teeth is the many species of Parrotfish found throughout the Pacific. There are also fish without teeth that have filters instead. These filters are used to suck in water and extract plankton from it. An example of a filter feeder is the largest fish in the world, the Whale Shark.

Markings

Markings, patterns and colors are distinctive and useful characteristics that can be used to identify fish by species, gender and developmental stage. Markings and patterns include spots, ocelated spots (a spot with a ring of another color around it), speckles (multiple fine spots), markings around the eyes (rings or patterns around the eye of the fish), blotches (irregularly shaped spots), vertical lines (bars), horizontal lines (stripes), and diagonal lines (bands). These markings, patterns and colors don't affect speed, movement or any other physical attributes, but they are used for other purposes. They are used to warn off other fish, for camouflage, and for mating purposes. For example a common Lionfish is brightly colored orange and red to warn predators of the poisonous spines along its dorsal fin. While Moray Eels use their color patterns as camouflage to hide from predators or so they can sneak up on prey.

IMPORTANT FISH SPECIES

The fish families that are being targeted by the GVI project because they are indicators of reef health are Surgeonfish (*Acanthuridae*), Unicorn fish (*Acanthuridae*), Moorish Idols (*Zanclidae*), Jacks/Trevallys (*Carangidae*), Sweetlips

(*Haemulidae*), Wrasses (*Labridae*), Emperors/Breams (*Lethrinidae*), Snappers (*Lutjanidae*), Goatfish (*Mullidae*), Mackerel (*Scombridae*), Groupers (*Serranidae*), Rabbitfish (*Siganidae*), and Barracudas (*Sphyraenidae*). Each of these families has their own subspecies. The most important of these species are discussed below. GVI targets specific species within each family, however other fish within these families can be surveyed as well. These fish are surveyed only by family and recorded as "Non-Target *family name*". Other important fish families are also targeted, but are surveyed only by family. These families include: Parrotfish (*Scaridae*), Butterflyfish (*Chaetodontidae*), Fusiliers (*Caesionidae*), Pufferfish (*Tetraodontidae*), and Triggerfish (*Balestidae*). The descriptions of these species were largely taken from the manual entitled *Reef Fish Identification for the Tropical Pacific* by Gerald Allen (2012) and from personal notes taken while surveying and studying these species in Fiji.

Surgeonfish

Surgeonfish, *Acanthuridae,* are reef fish whose bodies are often thin, oval shaped and have continuous dorsal and anal fins that run down the body. They also have crescent shaped tails and have small pointed mouths that they use to graze on algae, coral polyps, or consume zooplankton in midwater. Most have lateral lines and scales that are not prominent. Surgeonfish get their name from the scalpel sharp spine on each side of the fish at the base of the tail. This spine is used for fending off predators, defending territory, and establishing social dominance.

The juvenile Orangeband Surgeonfish, *Acanthurus olivaceus,* can grow up to 8 cm long and are initially completely yellow. They develop an elliptical orange band behind the top of the gill cover and change from yellow to light and dark gray as they age. The adult Orangeband Surgeonfish can be up to 35 cm long. The adult has lost its yellow colors and has a light gray head and front body and a dark gray second half. An elliptical orange band is located behind the top of the gill cover. They can live in groups or on their own. They live over sandy bottoms close to reefs at a depth of 3 to 45 meters.

The Striped Surgeonfish, *Acanthurus lineatus,* can grow up to 38 cm long. They are gold with many black edged blue stripes and a bluish belly. They have yellow ventral fins. All the other fins are blue with a light blue margin running through them. The spines on the tail are venomous and are used to defend against predators. It is mainly a solitary fish, but is sometimes known to live in groups with other Striped Surgeonfish. They live along the outer edges of the reef at a depth of about 6 meters.

The Yellowfin Surgeonfish, *Acanthurus xanthopterus,* grows to 56 cm. They have a blue or grayish brown body with yellow pectoral fins and have a yellow band running through the eye. This is the largest species of surgeonfish. They are both solitary or can form groups. They live in sandy areas near the reef at a depth of between 15-90 meters

The Lined Bristletooth, *Ctenochaetus striatus,* grows up to 26 cm. They are dark brown with thin blue lines throughout the body and small orange spots on the head. They can also have a small black dot at the base of the back of the dorsal fin. They live both in solitude or can form groups. They are one of the most abundant fish in lagoons and seaweed reefs and live at a depth up to 35 meters.

Unicornfish

Unicornfish are in the family of surgeonfish, *Acanthuridae,* but are in the genus *Naso.* Unicornfish are similar to Surgeonfish in terms of diet and have the same sort of flat, oval shaped body. There are physical differences between the two species that separate them. Their defining feature is not the unicorn horn that only some Unicornfish have, but the fact they have two prominent spines located on each side of their peduncle while Surgeonfish only have one spine per side. In general, Unicornfish are larger than the largest Surgeonfish and their bodies slope down to a very small peduncle. This is the best way to identify them, as it is not always possible to make out whether or not the fish have two spines.

The Bluespine Unicornfish, *Naso unicornis,* grows up to 70 cm. It has a gray or olive body with blue tail spines and a short forehead horn. The horn does not go past the mouth. It is a vegetarian and feeds primarily on leafy algae. It lives both in solitude or can form groups. They live in lagoons and on the outer reefs at between 1-80 meters.

The Orangespine Unicornfish, *Naso lituratus,* grows up to 30 cm. They have a brownish grey body with a yellowish nape and orange tail spines and anal fin. They have a yellow edged black area around the mouth up to the eye and a black band that runs along the dorsal fin. They live both in solitude or can form groups in lagoons and around the outer reefs at a depth up to 70 meters.

Moorish Idol

The Moorish Idol, *Zanclus cornutus,* grows up to 16 cm. and is the only species in its family. It has black, white, and yellow bars, an elongated white dorsal fin, and a long snout. It can be easily confused with the Longfin Bannerfish as they have similar colors, pattern and body shape. The recognizable yellow band on the snout and the black tail can differentiate the Moorish Idol from the Longfin Bannerfish.

They are omnivores, but mainly graze on algae, coral polyps and sponges. They live mainly in solitude but can be seen in pairs or small groups along coral reefs and lagoons to a depth of 180 meters.

Jacks/Trevallies

Jacks/Trevallies, *Carangidae*, are pelagic open water fish that are usually silver. They have flat torpedo shaped bodies and a sloping head with large eyes and mouth. The base of the tail is slender and the tail has a wide fork shape. Trevallies often travel in schools when going long distances. Their prey varies depending on the specific species of Trevally. All are carnivores. Most hunt smaller fish along the slopes of the outer reef.

The Yellow-Spotted Trevally, *Carangoides orthogrammus*, grows up to 70 cm long. It has yellow spots on its silver body. It has a blue fins and can sometimes have faint bars on both sides of the body. They can live in solitude or can form small schools. They live in lagoons over sand but can also be seen passing through and around outer reefs. They live at a depth of between 3-160 meters.

The Barcheek Trevally, *Carangoides plagiotaenia*, grows up to 42 cm long. It has a silver body with narrow dark bars on each of its gill covers. They live in solitude or can form small groups. They live along the edges and slopes of the outer reefs at between 2 and 200 meters.

The Bluefin Trevally, *Caranx melampygus*, grows up to 100 cm long. It has a silver body with an iridescent blue/green tinge to it. The top two-thirds of the body is covered with many dark and light spots. It lives in solitude or can form schools with other Bluefin Trevally. They live in many different reef habitats but mainly in the outer reef to a depth of 190 meters.

The Giant Trevally, *Caranx ignobilis,* grows to 165 cm. It has a silver body with many black dots along the entire body. There is a small black area around the base of the pectoral fin. This fish lives mainly in solitude in seaward reef slopes up to a depth of 80 meters.

Sweetlips

Sweetlips, *Haemulidae,* also known as Grunts, are very similar to Snappers in the sense that they have oval shaped bodies, triangular heads, emarginated tails and a continuous dorsal fin that is usually higher at the front than at the rear. The difference between the two is that Sweetlips are smaller than Snappers, have smaller mouths, more prominent lips and have flat teeth instead of canine-like teeth. They hunt at night for crustaceans at the bottom of the reef and spend the day in solitude or in groups around the reef. Juvenile Sweetlips often look very different from the adults.

Adult Oriental Sweetlips, *Plectorhinchus orientalis,* grow up to 85 cm. They have a white body with a yellow face and yellow fins/tail. Broad black stripes run along the body. There are broad black blotched patterns on the tail and fins. They live in solitude or can form small groups mainly in coastal reefs, seaward reefs and lagoons. They are a nocturnal predator on the reef, but swim in the open during the day. They live at a depth of between 2-25 meters.

Juvenile Oriental Sweetlips, *Plectorhinchus orientalis,* are between 4-8 cm long. They have a dark brown body with many large white spots surrounded by either an orange, yellow or red ring. They live mainly in solitude within the reef.

Sub adult Oriental Sweetlips, *Plectorhinchus orientalis,* are between 10-15 cm long. They have a white body with a yellow face and yellow fins/tail. They have broad black stripes that run throughout the head, body and fins. There are broad black blotched patterns on the tail and anal fins. They too live in solitude within the reef.

Wrasses

Wrasses, *Labridae,* are a family of reef fish that include 185 different species in the tropical Pacific. The body type, shape and size can be very different for each specific species, but every species of Wrasse has a terminal mouth, prominent canines, thick lips, and a continuous dorsal fin. Most are very colorful and can often be identified by their brightly patterned faces and large scales on their bodies. They are closely related to the parrotfish, but instead of eating algae they feed mainly on crabs, shrimp, brittle stars and small gastropods. They mainly use their pectoral fins for movement. Wrasses go through three phases of life, a juvenile phase, an initial phase, and a terminal phase. Wrasses are born hermaphroditic and when maturing to their terminal phase will often change genders to meet their reproductive needs.

The Terminal Phase Humphead Wrasse, *Cheilinus undulates,* grows up to 229 cm. It is also known as the Napoleon Wrasse or Maori Wrasse. It is a large fish with a green body with blueish tinge and has dark bars running down the body. It has a blue head with a green pattern on it. There is a large protruding hump right above the eyes of the fish. It lives primarily in solitude but can sometimes live in pairs. It lives in lagoons and the outer reefs to a depth of 60 meters.

The Initial Phase Humphead Wrasse is similar to the terminal phase in terms of overall shape and bluish green color. However, in the initial phase the hump above the eyes is not fully developed, but it is present. It can also be identified by the dark streaks around the eyes, a large diagonal steak going downward and two streaks starting at the back of the eye.

The Juvenile Humphead Wrasse is much different to the initial and terminal phases. It grows up to 20 cm and is gold yellow to yellowish green in color and has no present hump on its head. The head is triangular shaped and has two identifiable streaks starting from the back of the eye.

Other targeted Wrasse species include: the Tripletail Wrasse, *Cheilinus trilobatu;* the Floral Wrasse, *Cheilinus chlorourus;* and the Redbreasted Wrasse, *Cheilinus fasciatus.*

Snappers

Snappers, *Lutjanidae,* are a fairly large family of medium sized reef fish that have a sloped triangle shaped head, tapered body, a continuous dorsal fin, an emarginated tail, upturned snout, large mouths and canine like teeth near the front of both jaws. They are nocturnal predators that use their canines to mainly hunt smaller fish, but they also feed on gastropods and crustaceans. They live in shallow to medium depths along the reef.

The Five-Lined Snapper, *Luthanus quinquelineatus,* grow up to 39 cm. It has a yellow body, fins and head with five blue stripes running through the body. They are commonly found living in groups in coastal reefs, lagoons and outer reef slopes at depths of between 2-40 meters.

The Red Snapper, *Lutjanus bohar,* grows up to 75 cm. It has a red or reddish grey colored body. There is an indentation in front of the eyes of this species and the upper portion of the pectoral fin is a dark color. They live both in solitude or form groups in lagoons and outer reefs between 5-150 meters.

Other targeted Snapper species include: Humpback

Snapper, *Lutjanus gibbus;* Longspot Snapper, *Lutjanus fulviflamma;* Two-Spot Snapper, *Lutjanus biguttatus;* Onespot Snapper, *Lutjanus monostigma;* Mangrove Red Snapper, *Lutjanus argentimaculatus;* Blacktail Snapper, *Lutjanus fulvus;* Black-Banded Snapper, *Lutjanus semicinctu;* Midnight Snapper, *Macolor macularis;* and the Black Snapper, *Macolor niger.*

Emperor fish

Emperor fish, *Lethrinidae,* are a family of reef fish that are very similar to Snappers in terms of body shape and size. Emperors are medium to large sized fish with sloped triangle shapes heads, tapered bodies, a continuous dorsal fin, and an emarginated tail. They reside mainly along the edges of the reef. They are nocturnal hunters that prey mainly on sand dwelling invertebrates and sometimes on smaller fish. Many can switch their markings from having no pattern to having dark patterns, bars or spots on their bodies. The three main targeted species of Emperors include: the Thumbprint Emperor, *Lethrinus harak;* Orange-Stripe Emperor, *Lethrinus obsoletus;* and the Longface Emperor, *Lethrinus olivaceus.*

Coral Bream

Coral Bream, *Nemiperidae,* are a family of reef fish that are small to medium sized and reside along the sand surrounding the reef and near anything with structure (e.g., reef heads, rock gardens, shipwrecks). They have a sloping triangle shaped head, a tapered body, a small terminal mouth, and a continuous dorsal fin. They mainly live in solitude but have occasionally been known to form small groups.

The Humpnose Big Eye Bream, *Monotaxis grandoculis,*

grows up to 60 cm long. Their color ranges from black to grey or black to silver and they have a pale underbelly. There is often a yellowish tint around the head and a yellow lip. They live in solitude or form groups in coastal reefs, lagoons and the outer slopes of the reef to a depth of 100 meters. They are often confused with the Red Fin Bream as it was only in recent years that there was a distinction made between species. Some variations of the Humpnose Big Eye Bream have a dark top half of body with two thin silver bars dividing the body into thirds. This is very similar to the sub adult and adult phase of the Red Fin Bream and one of the reasons they are often confused. One of the main differences is the size. Humpnose Big Eye Breams generally have a thicker and overall larger body, while the Red Fin Bream is thinner and generally smaller. The other main difference is the tail shape. The Red Fin Bream has a more pronounced forked tail than the Humpnose Big Eye Bream.

Other targeted Bream species include: Rainbow Monocle Bream, *Scolopsis temporalis;* Bridled Monocle Bream, *Scolopsis bilineatus;* Longface Emperor, *Lethrinus olivaceus;* Redfin Bream, *Monotaxis heterodon;* and the Blue-Spotted Large Eye Bream, *Gymnocranius microdon.*

Goatfish

Goatfish, *Mullidae,* are a fairly large family of odd shaped reef fish. They are small to medium sized fish with a forked tail, two non-continuous dorsal fins and two barbells that protrude from the chin. These barbells are sensory appendages used to find worms, crustaceans, brittle stars and small fish. When not in use they are moved under the gill covers. Depending on the species, they can be either nocturnal or daytime hunters. Some Goatfish are able to change their colors and patterns while resting on the bottom for protection.

The Yellowstripe Goatfish, *Mulloidichthys flacolineatus*, grows up to 40 cm long. They have a silvery body with a yellow stripe running from just behind the head to the caudal peduncle. There is often a dark blotch on this yellow stripe, although not always. They are mostly seen schooling during the day and live on the reefs up to a depth of 76 meters.

Goldsaddle Goatfish, *Parupeneus cyclostomus,* grows up to 50 cm long. The most identifiable feature is the yellow saddle on their caudal peduncle. The body color can differ quite drastically from a dark blue, white, tan, brown, yellow or a mixture of colors. The yellow saddle is not visible on the all yellow version. They are often solitary but sometimes seen in pairs and live in reefs and lagoons up to a depth of 125 meters.

Other targeted species of Goatfish include: Yellowfin Goatfish, *Mulloidichthys vanicolensis;* Dash-Dot Goatfish, *Parupeneus barberinus;* Bicolor Goatfish, *Parupeneus barberinoides;* Doublebar Goatfish, *Parupeneus crassilabris;* Indian Goatfish. *Parupeneus indicu;* and the Manybar Goatfish, *Parupeneus multifasciatus.*

Mackerel

Mackerel, *Scombridae,* are a family of pelagic fish. They are larger fish and are often silver or grey in color but can have iridescent shades of blues and greens mixed in. Their long slender bodies are made for traveling long distances at high speeds. They have small dorsal and anal fins. They stay mainly in the pelagic region of the open ocean hunting squid and fish. The main targeted species of Mackerel in Fiji is the Narrow-Barred Spanish Mackerel. Although surveyed as non-target Mackerel, the most commonly seen species would be the Long Jaw Mackerel, *Rastrelliger kanagurta*. They are much smaller with a max length of 38 cm but are still an important food fish in the local markets. They can be seen in schools of hundreds feeding on plankton in mid water on reefs and drop offs.

The Narrow-Barred Spanish Mackerel, *Scomberomorus commerson,* grows up to 235 cm long. It is a long thin, pelagic silvery fish with thin chevron shaped bars along its body. A solitary fish, it can normally be found in open water but has been known to enter reefs, lagoons and swim along drop offs.

Groupers

Groupers, *Serranidae,* are a fairly large subgroup of the Sea Bass family. Depending on the species, a full grown Grouper can range in size from 40 cm to 231 cm. They have heavy bodies, continuous dorsal fins, complete lateral lines, large, wide mouths, multiple rows of teeth, round tail fins and three spines on each gill cover. Smaller species can mature to a point where they can reproduce within a year. Larger species can take many years to reach maturity where they can reproduce. Like many other organisms, mating is seasonal and controlled by the moon phases. Larger species travel to the place they were born to mate in mass schools. Groupers generally live in solitude near the bottom and feed on crustaceans and fish. Like many marine organisms, Groupers have a symbiotic relationship with cleaning shrimp and fish. Groupers are known to make frequent stops at cleaning stations along the reefs where they get cleaned of algae and parasites by shrimp or Gobys.

The Leopard Coral Grouper, *Plectropomus leopardus,* grows up to 70 cm long. The upper three quarters of the body is red and the bottom quarter is either grey, olive or dark brown. They have many small blue dots throughout the head, body, gill covers, and fins. They also have a distinct thin blue ring around their eye. They live in solitude along coastal and lagoon reefs at depths of 3-100 meters.

The Peacock Grouper, *Cephalopholis argus,* grows up to 60 cm long. Has an olive green or brown colored body with bright blue dots that have a dark rings around them. The tail and fins are a distinct dark blue regardless of body color. They are often found on their own hiding beneath branching corals or in small crevices at a depth up to 15 meters.

Other targeted species of Grouper include: Flagtail Grouper, *Cephalopholis urodeta;* Brown-Marbled Grouper, *Epinephelus fuscoguttatus;* Highfin Grouper, *Epinephelus maculatus;* Honeycombe Grouper, *Epinephelus merra;* Camouflage Grouper, *Epinephelus polyphekadion;* Squaretail Coral Grouper, *Plectropomus areolatus;* Giant Grouper, *Epinephelus lanceolatus;* and the Yellow-Edged Lyretail, *Variola albimarginata.*

Rabbitfish

Rabbitfish, *Siganidae,* are medium sized, thin bodied, oval shaped, colorful reef fish with small mouths, continuous lateral lines and venomous dorsal, anal and ventral spines. They are strictly herbivores and feed heavily on sea grasses and algae. The targeted species are: the Barred Rabbitfish, *Siganus doliatus,* and the Bicolor Rabbitfish, *Siganus uspi.*

Barracudas

Barracudas *(Sphyraenidae)* are large, long, silvery fish with a large, long mouth, large eyes, and large sharp teeth. They have a forked tail. The fins on a Barracuda are low, small and aerodynamic. These are the properties of a fast traveling pelagic predator. It feeds primarily on fish and has been known to eat young Barracuda. They can live in solitude but are commonly found in groups and schools.

The Great Barracuda, *Sphyraena barracuda, grows up to* 180 cm. They are the largest of the Barracudas. They have a long silver torpedo like body, with a forked tail, aerodynamic fins, a large elongated jaw and massive sharp teeth. There is a barred pattern along the sides of the body. They can live in solitude or form small groups and live mainly in reefs and in water up to 15 meters. Other targeted Barracuda species include the Blackfin Barracuda, *Sphyraena qenie* and the Pickhandle Barracuda, *Sphyraena jello.*

Family specific species

Parrotfish

Parrotfish, *Scaridae,* are a family of heavy bodied pectoral swimming fish. They are often brightly colored. They have a Juvenile Phase, an Initial Phase, and a Terminal Phase that they transition to as they get older. Some also have the ability to change gender from female to male when entering their Terminal Phase. Instead of teeth, parrotfish have hard bone fused beaks, which they use to eat algae, sea grass, and sponges. Some parrotfish take bites out of hard corals. Parrotfish often congregate and graze in schools. These schools are usually made up of a single alpha male and many females with whom they exclusively mate. While GVI surveys all Parrotfish by family rather than species, the exception is the Bumphead Parrotfish that is targeted for observation as an individual species because it is on the I.U.C.N. red list as being vulnerable.

The Terminal Phase Bumphead Parrotfish, *Bolbometopon muricatum,* grows up to 126 cm. It is the largest parrotfish.

It is green to olive gray in color, has a very prominent large bump on its forehead and a large plated beak. The Bumphead Parrotfish feeds on hard coral polyps. It uses its bump to break off large chunks of coral to make it easier to eat with its plated beak. They are slow moving and can often be seen in large herds. The sound of them grinding on coral can usually be heard from a distance before they are within sight. They live in lagoons and coral reefs to depths of 40 meters. The Juvenile Phase Bumphead Parrotfish is smaller and brownish green with five rows of white spots on each side.

Butterflyfish

Butterflyfish, *Chaetodontidae*, are small herbivorous fish with some species reaching a max size of 34 cm. Most stay within in 14-20 cm range. They are brightly colored, often swim in short bursts, and tend to school around coral heads. They are most often seen in and around branching corals.

Fusiliers

Fusiliers, *Caesionidae*, are small schooling fish that are often seen in the hundreds. They have slender torpedo shaped bodies, widely forked tails, and small mouths. Most have a stripe or stripe pattern running through their bodies. They are often seen feeding on plankton in mid water, on drop offs, or swimming in schools through coral reefs.

Triggerfish

Triggerfish, *Balistidae*, are medium sized predators with the largest species reaching up to 75 cm. They have a distinct diamond body shape and move around using primarily their anal and rear dorsal fin. The tail is only used in emergency situations. They have a distinct front dorsal spine that stands up straight like a "trigger" that in periods of danger serve as a warning sign to any organism that enters their territory. They are very territorial and large triggerfish have been known to bite divers and snorkelers. They have prominent teeth that are used to crush and eat shellfish and other invertebrates.

IMPORTANT INVERTEBRATE SPECIES

There are seven classes of invertebrates that the GVI project considers important. These include: Sea Urchins, *Echinoida;* Sea Cucumbers., *Holothuroidae;* Sea Star, *Asteroidea;* Bivalves, *Bivalvia;* Cephalopods, *Cephalopoda;* and Gastropods, *Gastropoda.* Within each class are certain species that are being targeted as indicators of reef health. The most important are described below. The descriptions of these species were largely taken from the manual entitled *Reef Creature Identification for the Tropical Pacific* by Gerald Allen (2012) and from personal notes taken while surveying and studying these species in Fiji.

Sea Urchins

Sea Urchins, *Echinoida,* are part of the Echinodermata phylum. Sea Urchins have a domed exoskeleton with spines. Some species have spines that deliver poison for protection. They have small tube like feet known as podia that are located underneath the animal that are used for movement. A mouth is located at the center of the bottom of the animal. The anus is located at the center on the top. They are mainly nocturnal feeders. Other targeted species of Sea Urchin include: Rock Boring Urchin, *Echinometra mathaei*; Small Black Urchin, *Echinostrephus aciculatu;* Coarse Spined Urchin, *Echinothrix diadema;* and the Striped Sea Urchin, *Tripneustes gratilla.*

Sea Cucumbers

Sea Cucumbers, *Holothuroidae,* are part of the Echinodermata phylum. Sea cucumbers have elongated cylinder shaped bodies with soft leathery skin. They have a mouth in the front, anus in the back, and small tubed feet either in their mouths or under their bodies. These small tubes are used for movement, holding onto the bottom, and for catching phytoplankton. These species are of a special interest because they are currently being over harvested to keep up with a growing demand from the Asian delicacy market.

An important species of sea cucumber is the White Teatfish, *Holothuria fuscogilva.* They grow up to 55 cm in length. They have a white elongated cylinder shaped body with soft leathery skin and brown or black splotches along the top. They live on sand or rubble bottom. Other targeted species of Sea Cucumber include: Curryfish, *Stichopus hermanni;* Greenfish, *Stichopus chloronotus;* Surf Redfish, *Actinopyga mauritiana;* Flowefish, *Bohadschia graeffei;* Leopardfish, *Bohadschia argus;* Brown Sandfish, *Bohadschia marmorata;* Lollyfish, *Holothuria atra;* Pink Lady, *Holothuria edulis;* Elephant Trunkfish *Holothuria fuscopunctata;* Black Teatfish, *Holothuria whitmae;* Sandfish, *Holothuria scabra;* Prickly Redfish, *Thelenota ananas;* and the Megabum, *Thelenota anax.*

Sea Stars

Sea Stars, *Asteroidea,* are part of the Echinodermata phylum. Depending on the species, Sea Stars have five or more arms protruding from a central disc. There are two or four rows of suction cup podia that line the bottom of each arm from base to tip. The central disc has a mouth on the bottom side and an anus on the top. If a limb is severed, Sea

Stars have the ability to grow it back. The most important targeted sea star is the Crown of Thorns Sea Star, *Acanthaster planci.* They can grow up to 40 cm. They are a large sea star that is covered with sharp poisonous thorns that are painful but not toxic. Colors are wide ranging, including red, purple, yellow and tan. They feed on coral polyps. They can live in solitude, but when there is a population explosion they have been known to form groups that can decimate large chunks of coral reef. However, when populations are in healthy numbers the Crown of Thorns Sea Star plays an important role in keeping up the biodiversity of corals as they primarily prey on fast growing branching corals allowing slow growing bordering corals to compete.

Bivalves

Bivalves, *Bivalvia, are* part of the Mollusca phylum. Bivalves have soft bodies that are encased and protected by two shells. A strong adductor muscle holds these shells together. The shells can vary greatly in terms of size, color, shape, pattern and texture. Bivalves detect changes in light using eyespots along the rim of the opening. The bivalves targeted for survey by GVI feed using suspension feeding. Suspension feeding is the process of sucking water through a filter and feeding off the phytoplankton or other organic material that is captured in the filter.

A targeted Bivalve in Fiji is the Giant Clam, *Tridacna spp.* It can grow up to 130 cm. This is the largest bivalve in the world. The two outer shells have four folded ridges where most of the color and pattern is. The opening between the two shells is fused. There are two holes in opening which are used to siphon water in and out of the clam. The giant clam is an omnivore and will eat anything that is in the water around it but mainly feeds on phytoplankton. They live in the reef or in lagoons.

Cephalopods

Cephalopods, *Cephalopoda,* are part of the Mollusca phylum. Cephalopods have eight arms/tentacles that are equipped with suction cups that they use to capture prey and sharp beaks that they use to consume or kill their prey. They have no external shell or exoskeleton, instead they have soft bodies. They are able to avoid precarious situations by either using a sudden jet of water to escape quickly, excreting an ink cloud, or by using camouflage. Octopus and Cuttlefish are known for their ability to change the color of their skin to match their surroundings. A common Cephalopods in Fiji is the Common Octopus, *Octopus cyanea.*

Gastropods

Gastropods *(Gastropoda)* are the largest class of the mollusk phylum. They typically have a soft body and a hard protective shell. The shell is what is mainly used for identification of these creatures. The shells vary widely in terms of size, color and shape. They have two tentacles with eyes on the end that are used for vision. They feed primarily on plants or algae.

An important Gastropod in Fiji is the Triton's Trumpet, *Charonia tritonis.* They can grow up to 50 cm. They have a long cone shaped spiral shell with white and brown bands. The soft body is also white with brown splotches and they have alternating yellow and black banded antennas. They are one of the only natural predators of the Crown of Thorn Sea Stars. Other targeted species of Gastropods include Drupella, *Drupella cornis;* Coral Snail, *Coralliopjila neritoidea;* and the Tiger Cowrie, *Cypraea tigris.*

IMPORTANT BENTHIC SPECIES

Benthic organisms are the easiest of the marine species to observe because unlike fish and invertebrates they don't move. There are 19 species of coral and plants that are being targeted for observation in Fiji. These 19 species can be split into three categories, Non-Acropora, Acropora and Other Fauna. Hard corals come in many different shapes. The two categories of hard corals, Non-Acropora and Acropora, are categorized by the presence of terminal polyps on their branches. Because Acopora corals are required to have terminal polyps they can only come in three shapes. These include:

- Acropora Branching corals with at least two degrees of branching with terminal polyps at the end.
- Acropora Digitate corals with one degree of branching with terminal polyps at the end.
- Acropora Tabular corals - an Acropora Branching coral that forms a low flat table shape that originates from a central stem. Small Tabular corals are often seen growing off of larger corals heads, while larger Tabular corals are often seen growing off of the bottom, i.e., sand or rubble.

All other hard corals that don't have terminal polyps are classified as Non-Acropora. These hard corals take shapes that include:

- Non-Acropora Branching corals with at least two degrees of branching, but with no observable terminal polyps.

- Non-Acropora Encrusting corals that grow and encrust objects including other coral, dead coral, rocks, sea fans or ship wrecks.
- Non-Acropora Foliose corals - thin leafy corals that are often over lapping with each other forming cabbage like corals shapes.
- Non-Acropora Mushroom corals that form small mounds that can become elongated mounds. Unlike other corals, mushroom corals are not anchored to the substrate and are often seen flipped over.
- Non-Acropora Tubipora coral looks like they have small white flowers on soft stems growing out of it. These flowers move when wafted and this gives off the impression that it is not a hard coral, however when wafted continuously the flowers retract revealing the hard coral skeleton.
- Non-Acropora Massive corals are a boulder or mound shape coral that can be small or large. The name has nothing to do with size.
- Non-Acropora Submassive corals are corals that form shapes that do not fall into any of the other categories. They have columns or jagged shapes. This includes growth from encrusting corals that grow on original encrustings and create their own free form shape.
- Millepora corals are commonly known as fire coral. They are identifiable by their orange rust color with the small white tips at the end of its branches and the small hairs that can be seen projecting off of the coral. It is most often seen as a Non-Acropora Branching coral, but can also be seen as an encrusting coral, submassive coral, and a submassive coral growing out of an encrusting coral.

Hard corals and reef building corals have a hard exoskeleton made up of calcium carbonate created by live coral. Live coral polyps live atop the skeleton of older corals. This process takes place over thousands of years and is responsible for the coral reefs of today. The color of the corals come from zooxanthellae, a single cell algae that coral has a symbiotic relationship with. The corals host the algae and in return the algae produce energy for the coral through photosynthesis. While they derive a large portion of their energy through photosynthesis, corals are not autotrophic. Corals also feed on phytoplankton using small tentacles on each of the coral's polyps. When these corals are actively feeding it may look as if small wafting tentacles have extended off the hard coral body. This may confuse surveyors studying hard corals into thinking it is an anemone or a soft coral. Another point of confusion when surveying corals comes from the fact that the corals are surveyed by life form rather than species. A single head of coral can have multiple life forms. In order to deal with this, the surveyor surveys only what falls underneath the transect tape. If the coral head changes life form underneath the tape, the surveyor records each change. This is most common in encrusting corals. An encrusting coral that completely encases a rock becomes a massive life form. However, if a free form shape were to grow off of the massive structure that part of the coral would be surveyed as a sub massive life form. The surveyor would record each change in life form as it comes underneath the transect tape.

The other benthic organisms that are targeted for observation fall into either the "Other Fauna" category or the "Algae" category. This "other fauna" category consists of non-hard corals like, Soft Corals, Sponges, Zoanthids, Tunicates, Anemones, Giant Clams, and Gorgonians, like

Sea Whips and Sea Fans. The algae category contains three different types of targeted algae, Macro Algae, Turf Algae and Crustose Coraline algae.

Fiji is famous for its abundances and diversity of Soft Corals. They are often found in areas that are heavily affected by tides (i.e., close to shore where the high and low tide is very visible) and in areas where there is a strong current (i.e., deep channels and along coral walls). Soft corals use the movement of water to their advantage and feed heavily on the phytoplankton that pass by. When a current or movement of water is present, these soft corals appear to "bloom" as they expand and release their polyps' tentacles to feed. Soft corals are softer and more flexible than hard corals, hence the name, and are found moving with the currents. Because they are flexible, they can withstand strong currents that would otherwise damage them. However, soft corals are fragile in the sense that they are very vulnerable to changes in water temperature and pH. Like hard corals, soft corals come in different shapes and varieties. These include: Carnation Coral (*Dendronephthya*), Toadstool Coral (*Sarcophyton*), and Tree Corals (Nepthidae).

Macro algae is the most common type of algae. It comes in many different forms. Most forms are greenish yellow to green. They can take the form of leafy algae, disc algae or can look like a shrubbery algae. Turf algae is a greenish brown algae that resembles small wispy hairs that can often be seen growing off of dead corals or bleached corals. Crustose Coraline Algae is a pink or red algae that resembles an encrusting coral, yet when touched it is slimy. It acts as cement that binds broken corals and rubble together. It is important in the rebuilding of a coral reef after a natural disaster.

I.U.C.N. SPECIES

Many of the species that GVI targets on both normal surveys and opportunistic surveys are on the I.U.C.N. Red List. These were described above. Other species on the I.U.C.N. Red List not yet described include: Requiem Sharks, *Carcharhinidae;* Nurse Sharks, *Ginglymostomatidae;* Stingrays, *Dasyatidae;* Eagle Rays, *Myliobatidae;* and Sea Turtle, *Cheloniidae.* The descriptions below of these organisms come largely from the manuals entitled *Reef Creature Identification for the Tropical Pacific* by Gerald Allen (2012) and *Reef Fish Identification for the Tropical Pacific* by Gerald Allen (2012), and from personal notes taken while surveying and studying these species in Fiji.

Requiem Sharks, *Carcharhinidae*

Requiem Sharks have the stereotypical shark shape. There are 48 different species of Requiem Sharks. These sharks are streamlined and powerful with a large dorsal fin, pointed snouts, distinct pectoral fins, and a tail with a large upper lobe and smaller lower lobe. The main species that were observed were Bull Sharks (*Carcharhinus leucas*), Tiger Sharks (*Galeocerdo cuvier*), Lemon Sharks (*Negaprion acutidens*), Silvertip Shark (*Carcharhinus albimarginatus*), Blacktip Reef Shark (*Carcharhinus limbatus*), and Whitetip Reef Shark (*Triaenodon obesus*).

The Bull Shark is a heavy bodied bulky shark that can grow up to 3.4 meters. It has a dark gray upper half of body that fades to a white bottom half. The bulky body slopes down to a blunt snout. They are considered to be the most dangerous shark in the world because they are responsible for more shark attacks on humans than any other shark. However, this can be attributed to the fact that Bull Sharks, unlike other sharks, can be found in tropical water, colder temperate water, salt water, brackish water, and freshwater. Therefore, the chances of running into this shark are much higher than any other shark which results in more chances for shark attacks. Bull sharks spend the majority of their lives in open water and on reefs. Each year, mature bull sharks travel and congregate in traditional mating grounds. They give birth in the same fresh water river that they were born in. Bull Sharks can live in depths of up to 152 meters.

Tiger Sharks are a large shark that can grow up to 5.5 meters. They have distinctive grayish white stripes along the side of the body, a very large head and mouth, and an elongated tail. Tiger Sharks are a very large shark that is thought to be very dangerous, yet little is really known about their behavior. They are not a schooling shark and prefer to be on their own. They are not thought to be a localized shark and often swim great distances. They can live in depths up to 75 meters.

Lemon Sharks can grow up to 3.1 meters. They have distinctive long sharp teeth that sometimes stick out of their mouth. They have a small pointed head and the mouth seems as if it is smiling. It has two dorsal fins that are the same size. They are commonly found swimming along the bottom in water between one and 30 meters.

The other shark species commonly observed in Fiji is the Tawny Nurse Shark (*Nebrius ferrugineus*). Unlike most sharks, this shark feeds using powerful suction and mainly feeds on crustaceans. It is brownish gray in color and has no large visible teeth. It has two barbells that it uses to detect prey. They can grow up to 3.2 meters and live at depths between one and 70 meters.

OBSERVATIONS FROM FIJI

Through classroom training sessions, teaching in the local schools, interacting with local community members, and while surveying the species listed above, lessons were learned about the importance of these species as indicators of reef health, the important role these species play in maintaining the overall health of the reef, their importance as a local food resource, the commercial importance of these species, as well as the threats that are currently faced by these species. From the smallest of Butterfly Fish and Sea Urchins to the largest of Acropora Branching corals and Humphead Wrasse, each species is being surveyed for a reason. The importance of each species is often the same for the species' families, yet certain species will also have individual importance. Listed below are some of the key facts learned while in Fiji.

OBSERVATIONS OF SURVEYED FISH SPECIES

Algae grazers: Butterflyfish, Moorish Idols, Surgeonfish, Unicornfish, Rabbitfish and Parrotfish.

A healthy abundance of these types of fish are a good indicator of reef health. Algae grazers feed on different types of algae; e.g., Macro Algae, Turf Algae, and Crustose Coraline algae. These types of fish feed mainly on algae that is growing on the bottom and on Zooplankton found in mid to open water. Healthy populations of algae grazers are required to counterbalance large amounts of algae that can smother coral. An abundance of these types of fish that are found on the reef is often linked with a healthy amount of algae in the area or a recent growth of algae. Too much algae is known to smother and outcompete the slower growing corals that make of up the reef.

A school of Butterflyfish, Moorish Idols, Surgeonfish, Unicornfish, or Rabbitfish in mid water or open water is often a sign of a high amount of Zooplankton in the water. The main threats to these fish are overfishing and habitat loss. Overfishing of algae grazers results in uncontrolled growth of algae which destroys the habitat for the majority of reef fish and commercially important fish, especially juveniles which use corals for shelter and protection.

For example, a cyclone badly damaged Caqalai in 2010, destroying or damaging many of the hard corals found around the house reef. Due to the destruction of hard corals and other factors along the windward side of Caqalai, large amounts of Macro, Turf and Crustose Coralline algae began to grow. This new growth of algae has caused an increase in the population of algae grazers along that side of the island.

Coral grazers: Parrotfish

Like the other algae grazers, parrotfish play an important part in maintaining a balance between algae and coral growth. However, parrotfish are unique in that they are also able to feed on hard corals. The other algae grazers only feed on leafy algae. This allows parrotfish to graze on algae that could be growing over hard corals. Although Parrotfish are not a typical commercial food fish like Snappers, Groupers or Tuna, they are subject to overfishing and can often be found in local fish markets.

The Bumphead Parrotfish is known for its large bump, large plated beak and being able to grow over a meter in length. It is currently listed on the I.U.C.N. Red List as vulnerable to extinction. Due to its large size it was prized for food in many cultures, but because of its slow reproduction rate and slow gentle moving nature it has been almost hunted to extinction in Fiji.

Commercially important/Food fish: Snappers, Groupers, Emperors, Tunas, Mackerel, Barracuda, Sweetlips and Trevally

These fish are all middle to high level predators in the food web. They are predators of smaller fish, crustaceans and other invertebrates. They themselves are prey of higher level predators like sharks, bigger fish and humans. A site containing a healthy population of these types of fish is often found to be in overall good health. A healthy population of these types of fish acts as a counter-balance in the ecosystem. They make sure the population levels of smaller fish, crustaceans and other invertebrates do not get too high. Their populations are kept in balance because they are a food source for sharks, bigger fish and humans.

For centuries this group of fish has been a staple food source of island nations and coastal populations. Without these fish a staggering amount of people would lose an important source of protein. As human population in coastal communities has risen, so has the demand for these fish. Combined with the industrialization of the fishing industry (i.e., long lines, massive trawling nets and deep sea fishing), populations of these fish have come under heavy pressure. The World Wildlife Fund states, "unless the current situation improves, stocks of all species currently fished for food are predicted to collapse by 2048" (Worm, 2006). These fish are also vulnerable to habitat destruction due to coral reef loss because of low algae grazer populations, nutrient run off from farms, factories or coastal hotels, climate change, natural disasters, or the destruction of mangroves from coastal development.

Because the largest of the fish are being caught for commercial export, local fishermen are now finding themselves catching move juvenile and smaller fish than what they used to. This results in a loss of mature fish, fish that are mature enough to breed and reproduce, and a loss of juvenile fish that have yet to even reach breeding age.

Other important species: Fusilers, Triggerfish, Goatfish, and Wrasses

Fusilers are small schooling fish that can be found in large schools containing hundreds of individuals swimming in mid to open water. They feed primarily on Zooplankton drifting in the water. They are hunted by larger fish like tunas, mackerels, and barracudas. Their role in the marine ecosystem bridges the gap between autotrophic plankton and the larger predators in the food web. An abundance of Fusilers indicates a healthy amount of zooplankton in the water. A shortage of zooplankton would inhibit energy transfer from the sun to the surrounding reef and negatively affect all the organisms that live there. The biggest threat to Fusilers is climate change; more specifically, ocean acidification, which has serious negative effects on zooplankton, the primary source of energy for Fusilers.

Triggerfish, Goatfish, Breams and Wrasses are all small to medium sized reef fish. These four families primarily feed on crustaceans and invertebrates and help keep a balance in those populations. A healthy population of these fish indicates a good balance between the crustaceans like shrimp, crabs, mollusks and the reef fish. Many of these fish species are commonly hunted and eaten by local fishermen, but are not generally caught for commercial purposes. However they can still be found in local fish markets around Fiji. These fish that breed, hunt and live on the reef have habitat destruction as their biggest threat – be it from natural disasters, overfishing, coral bleaching due to climate change, pollution, or invasive species. Some of these species, especially some of the Wrasses, have their own important role in maintaining reef health.

For example, the Humphead Wrasse plays an extremely important role in the maintenance of reef health. They are one of the only natural predators of the Crown of Thorns sea star. An invertebrate that is notorious among fisherman and conservationists as being destructive to reefs when populations get too high. The Crown of Thorns sea star is extremely efficient at preying on live corals and have been known to wipe out large portions of a reef. The Humphead Wrasse regulates the population of these Sea Stars keeping them at healthy, non-destructive levels. Unfortunately, through overfishing and habitat loss, the Humphead Wrasse is now listed as endangered on the I.U.C.N. Red List. This may be the cause of more frequent population explosions of Crown of Thorn sea stars. Even though this fish is listed as endangered, local fishermen are still catching them. Many fishermen know about the dangers posed by the Crown of Thorns sea star, but are unaware of the importance of Humphead Wrasse in regulating their populations.

Another example is the role of the Cleaning Wrasse. Though Cleaning Wrasse are surveyed by GVI as Non-Target Wrasse, they still play an extremely important role in the reef ecosystem. They are small, no longer than 5 cm, slender fish that live within the reef on "cleaning stations." They have a symbiotic relationship with many of the fish that live within the coral reef that visit these cleaning stations. These can include large fish like Snappers or Groupers and smaller fish like Butterflyfish or Surgeonfish. As the name implies, the Cleaning Wrasse clean these fish by feeding on the parasites, algae or other material that are stuck on the visiting fish. They do not have many predators as they perform an important function for those fish that might otherwise prey on them. Cleaning Wrasse can often be seen swimming in and out of larger fishes' mouths. Cleaning stations with good populations of Cleaning Wrasse indicate a healthy population of resident reef fish in the area. Because these Wrasse live within the reef and rely on their symbiotic relationship with the larger fish for food, the biggest threats facing them are overfishing and habitat loss.

OBSERVATIONS OF SURVEYED INVERTEBRATE SPECIES

Algae grazers: Sea Urchins, Spider Conch, Trochus, Tiger Cowrie (juvenile)

Like their fish counterparts, a healthy population of algae grazing invertebrates is a good indicator of reef healthy. These algae grazers feed on different types of algae; e.g. Macro algae, Turf algae, and Crustose Coraline algae. A healthy population of these grazers acts as a regulator for the fast growing algae and prevents them from overgrowing and smothering the slow growing corals. Most of these invertebrates are the prey of Triggerfish, Goatfish, Breams and Wrasses. In this way they facilitate the transfer of energy from autotrophic algae higher up in the food web.

The colorful shells of the Tiger Cowrie and Spider Conch are commercially important in the shell collection and tourism trade. Unfortunately, due to an unsustainable demand for shells, poaching has caused the populations of these invertebrates to drop dramatically. This dramatic drop in algae grazer populations has resulted in the unregulated and expansive growth of algae in some areas and the smothering of many corals.

Coral predators: Crown of Thorns Sea Star, Drupella, Tiger Cowrie (adult)

This infamous group of invertebrates feeds primarily on live coral polyps that are found in hard corals. While these coral predators are from different biological families, they feed in essentially the same manner. They expel their stomachs over the hard corals and use stomach acids to dissolve the hard coral skeleton and digest the live coral polyps living within. This digestion leaves a bleached white "scar" mark similar to coral bleached by solar radiation and high ocean temperatures. This bleached section of the coral is vulnerable to growth of algae and can act as a starting point for fast growing algae to expand over the coral. When these invertebrate populations are not too high they provide an important service to the health of the reef. Like any other predator-prey relationship in the food web, these invertebrates regulate the biological diversity of their prey. This results in a healthy diversity of coral species on the reef and ensures that a faster growing species does not overgrow and take over the reef. Unfortunately in recent years with the decline of the predators that feed on these invertebrates (i.e., the Humphead Wrasse, Triton's Trumpet, and Triggerfish), the populations of these coral predators are becoming too abundant. These uncontrolled increases in population have caused serious damage to coral reefs around the world. An area that is found to have a large population of these coral predators can often be correlated with overfishing of their natural predators.

These species, especially the Crown of Thorns sea star, are infamous around the world and the damage they can bring to a reef is widely known by scientists, politicians and local fishermen. However, it seems, that many fishermen do not know who the natural predators are and their importance in controlling the population of these coral predators.

Predator of coral predators: Triton's Trumpet

The Triton's Trumpet is a large invertebrate with a beautifully patterned horn shaped shell. It is one of the few organisms that feed on the Crown of Thorns sea star. This makes it an extremely important species in the coral reef ecosystem. Along with the Humphead Wrasse, this species regulates the population of the Crown of Thorns sea star. Healthy populations of the Triton's Trumpet on coral reefs are a good indicator of controlled populations of the Crown of Thorns sea star. Unfortunately, due to its decorative shell, it is under heavy demand in the seashell trade. For many years indigenous people around the Pacific have used it as an air horn, now local communities collect them mainly to sell to tourists to use as decoration. With populations of Triton's Trumpet falling because of over-collection and populations of Humphead Wrasse falling because of over fishing, the population of Crown of Thorns sea stars is growing. Unless something is done to protect these predators, population explosions of the Crown of Thorns sea star are going to happen at a more frequent rate. To repeat what was said above, many members of local communities who collect this invertebrate either do not know that it is a natural predator of the Crown of Thorns sea star or do not understand its importance in curbing the danger posed by the Crown of Thorns sea star. This is quickly becoming one of the biggest threats to coral reef health in the Pacific and will get much worse if something is not done soon to fix the problem.

Sediment filters: Sea Cucumbers

There are 14 different species of Sea Cucumbers that GVI surveys. Although they may not look important, all 14 of these species play a critically important role in the maintenance of reef health. Sea Cucumbers are the reef's natural filters and cleaners. They filter the sand and water and feed on decaying matter, fish and invertebrate waste and pollutants. Most importantly these species filter out excessive nutrients due to run-off from farms, sewage plants, cities, factories, and hotels. They can often be found around places where that run-off flows into the oceans (i.e., end of sewage pipes). An abundance of Sea Cucumbers often indicates a high amount of nutrients in the area. Their role as filters and cleaners also make them regulators of the amount of fast growing algae in the area (MacTavish, et al., 2012). Fast growing algae thrive in nutrient rich sand and water. These conditions often lead to algae blooms in which corals are often overgrown and smothered. This can result in major habitat loss for many fish and invertebrates that rely on coral reefs for shelter and food.

Due to a rising demand from Asian countries, Sea Cucumbers are being overfished to dangerous levels. The Asian delicacy Beche-de-mer, that is dried or cooked Sea Cucumber, is currently in heavy demand. Asian fishing companies are hiring local fishermen to collect Sea Cucumbers and paying thousands of U.S. dollars. This large amount of income is almost unheard of in the small fishing villages on some of Fiji's more remote islands and is bringing in needed money to build infrastructure and purchase boats, fishing nets and refrigeration equipment. Unfortunately these monies allow local fishermen to buy more industrialized fishing equipment and with their new ability to sell commercially, these fishermen are overfishing the already depleted reef fish populations. Due to this ability to realize higher income from Sea Cucumbers, their populations on the shallower parts of the reefs are being depleted. With the demand still high, fishermen are diving deeper in order to collect them and Asian companies are supporting this. They are supplying these fishermen with boats and scuba gear, but with no safety training. This has resulted in a large increase in the number of diving related injuries from decompression sickness, lung overexpansion, and drowning. Overall, these invertebrates are both biologically important and commercially important, but an unsustainable demand is putting their populations at levels that are detrimental to overall reef health.

OBSERVATIONS OF SURVEYED BENTHIC SPECIES

Algae: Macro Algae, Turf Algae, Crustose Coraline Algae

As was stated above, too much algae is never good for reef health. However, algae are still a very important part of a healthy reef ecosystem. As autotrophic organisms, algae use sunlight and nutrients in the water to create sugars and energy. Along with coral polyps and plankton, algae make up the base of the food web and are the beginning of the energy transfer process. Fast growing algae feed many species of reef fish and invertebrates, primarily algae grazers (e.g., Butterflyfish, Surgeonfish, Trochus) These algae are also good at regulating excessive amount of nutrients in the water. Yet, when there is an excessive amount of nutrients in the water stemming from farms, factories, hotels, etc., algae populations explode. These large increases in abundance of algae are called algae blooms and can be very detrimental to reef health. However, constructively, large amounts of Crustose Coraline Algae can benefit coral reefs that were damaged by cyclones or other natural disasters. Crustose Coraline Algae is similar to encrusting coral in that it grows over exposed rock or dead corals. In doing so it acts as a binder and holds together pieces of dead or broken coral. This begins and speeds up the process of rebuilding the reef.

As was stated above, because algae grows faster than corals, algae blooms can outgrow and smother live hard corals that make up the structure of the reef. Large abundances of algae can be caused by overfishing or over-collection of algae grazers and predators of coral predators, an excessive amount of nutrients found in the water, and/or recent coral damage from cyclones and natural disasters. That being said, when algae populations are kept under control by algae grazers, they provide an important role in the overall health of the reef.

Hard Corals: Acropora and Non Acropora corals

The 11 different types of hard corals that are surveyed are the basis of the coral reef ecosystem. The hard calcium carbonate skeletons of the hard corals are what give coral reefs their structure. The coral polyps that contain symbiotic Zooxanthellae that live in the coral skeleton slowly form these hard calcium carbonate structures. These polyps slowly release calcium carbonate as they continue to expand. The calcium carbonate is accumulated from minerals found in the water and the required energy stems from photosynthesis and zooplankton these polyps feed on.

These coral skeletons provide shelter from predators, hunting grounds for predators, and nurseries for many different species of fish and invertebrates. The huge abundance of species living among the coral provide a primary food source and therefore financial living to people living in neighboring communities. Many species of juvenile fish that are caught for commercial purposes when they mature live on the reef until they are large enough to wander. From the tiniest of coastal villages to large cities, millions of people rely directly or indirectly on the fish from these reefs. They also protect the coastline and beaches from incoming storms and natural disasters by acting as a barrier for incoming wind and wave action.

Unfortunately these corals are very fragile, easily stressed, and require very specific conditions to be able to thrive. There are many threats to this extremely important part of the marine ecosystem. When stressed, they expel their Zooxanthellae and are put in a situation where death is likely. Corals that are exposed to a temperature rise of three to four degrees Celsius for one or two days or a rise of one to two degrees Celsius for a few weeks are susceptible to coral bleaching and often death (Jokiel & Coles, 1990). A large presence of coral predators such as the Crown of Thorns sea star or Drupella can also cause a high amount of stress that leads to coral bleaching and coral death. A high amount of nutrients combined with coral bleaching also leads to a high probability of an algae bloom. In addition, Macro and Turf algae that thrive under conditions where there is a high level of nutrients in the water, can grow rapidly over the calcium carbonate skeleton of bleached coral when there are no Zooxanthellae present to compete with the algae.

The destruction of mangrove forests for coastal development also adds to this problem. Mangrove forests act as natural filters and barriers to sediment, pollution and nutrient run off from farms, factories and coastal hotels. Clearing of mangrove forests for development allows high levels of run-off to create conditions that algaes thrive in and allow for more frequent algae blooms. Mangroves also act as a natural nursery for many algae grazing fish, such as Surgeonfish and Butterflyfish. Destruction of these nurseries results in a population decline of the regulators of algae populations in the same way that over-fishing and over-collection of algae grazers pose a threat to the health of these corals.

In the face of all these threats to hard corals, an abundance of healthy corals indicate things are balanced in the ecosystem. Healthy corals can indicate a healthy population of algae grazers, which can be a result of habitat protection and an absence of over-fishing. A healthy abundance of corals also means that there is a controlled population of coral predators, which means that predators like the Humphead Wrasse and Triton's Trumpet are protected from overfishing, over collection and habitat loss. Corals need a specific temperature range in order to thrive. Hence, a healthy population of corals would indicate that temperatures are not abnormally fluctuating and are not getting too hot. Observing a healthy population of hard corals is an excellent sign that things in the marine ecosystem are balanced and healthy.

OBSERVATIONS OF I.U.C.N. SPECIES

Top level predators: Tiger Sharks, Bull Sharks, Lemon Sharks, Tawny Nurse Sharks, Silvertip Sharks, White tip and Black tip Reef Sharks.

In the same way that algae grazers regulate algae populations and the Humphead Wrasse regulates Crown of Thorn sea star populations, these top level predators regulate the populations of medium to high level predators such as Tuna, Snappers, Rays, and Turtles. This regulation trickles down the food web. When the high level predators are kept at a healthy population level, the medium level fish and invertebrates are kept at a healthy population level. When those predators are at a healthy population level, the algae grazers are kept at a healthy population level and so on. It is all linked. These top predators stop other predator population levels from getting too large and wiping out the species they prey on. In order to keep sharks and other top level predator populations from getting too large, they often have long gestation periods of a year or more. They also take a long time to reach sexual maturity. For instance, Bull Sharks do not start reproducing until they are in their teenage years.

Despite their size, most sharks are stealthy hunters and often hunt in poor visibility or at night in order to sneak up on their prey. For instance Bull Sharks will sneak up on unsuspecting mammals or birds that are swimming in muddy rivers. However, sharks prefer animals that are already dead or prey that is already injured as they have a much higher chance to catch it.

The majority of sharks will steer clear of humans as they are too big to eat and do not resemble a shark's typical food. The large majority of shark attacks on humans are a result of bad visibility, confusion or starvation. Most shark attacks are only quick test bites as that is the only way they can test if something is food or not. Unfortunately, if a five meter tiger shark takes a quick test bite it will often result in the loss of a limb or in a loss of a life.

Humans are what threaten shark populations. The most blatant threat to sharks is the current demand for shark fin soup. This traditional Asian delicacy has risen in popularity in the past 30 years and is commonly served at Chinese weddings and important meetings. However, the actual shark fin has little to no taste and is only used as a superficial status symbol. The demand for shark fins has resulted in the rise of a barbaric and destructive fishing practice called "shark finning" where a shark is caught and while the shark is still alive has its dorsal, anal, pectoral, and tail fins sliced off. The shark is often thrown back in the water to make room for more shark fins. This practice has been condemned around the world as a destructive practice but is still legal in many countries including Fiji. WWF reports that at least 100 million sharks are killed each year through fishing, by-catch, or shark finning (WWF, 2015). These numbers are totally unsustainable and have resulted in the majority of shark species being listed on the I.U.C.N. Red List. Like rhino horn, lion bones and elephant tusk, the traditional Chinese medicine market also claims that dried shark fin is a miracle cure for cancer, improves sexual performance, and other such nonsense. The miracle qualities of shark fin have been disproven. In addition, other serious man-made threats to sharks include being caught as by-catch in large industrial fishing nets and long lining vessels and the direct and indirect effects of habitat destruction of mangrove forests and coral reefs.

However, the biggest threat to shark species is not overfishing of sharks, but overfishing the food that sharks traditionally feed on. Due to the industrialization of the fishing industry and the use of large trawling nets and longlines, populations of many of the traditional fish that these sharks prey on are dwindling. The lack of available food is causing sharks to either starve or move to non-traditional areas in search for foods. This is one of the reasons for the increase in shark attacks in recent years. Sharks are venturing into waters where there has never been a recorded shark sighting in search of food. Due to their uncertainty being in an unknown place and the forced hunger of these sharks, they start taking more test bites to see what is and is not food. Unfortunately many of these new areas are places where humans are found. The more our oceans are overfished, the more frequent shark attacks are likely to happen.

COMMUNITY BASED MARINE CONSERVATION

INTRODUCTION

Community Based Marine Resource Management (CBMRM) is the use of Community Based Natural Resource Management techniques that have been adapted to conserving marine resources. CBMRM strategies are used by Non-government Organizations (NGOs) like the World Wildlife Fund (WWF) and government agencies like the United States Agency of International Development (USAID). These and many similar groups worldwide work with communities to help conserve marine resources that communities rely on for their welfare. Projects using CBMRM techniques work with communities to give them the knowledge and skills to better manage their environment. CBMRM techniques may also help communities develop rules and regulations to control the use of a resource; for example, setting a size limit or banning all together the catching of a threatened species to help the species recover. Rules and regulations can also be used to control how an area is used; for example, banning fishing in a protected reserve, but still allowing scuba diving there to bring income into the community through tourism.

CBMRM projects are based on existing systems of government in these communities. CBMRM involves respecting and working with pre-existing land rights, chiefdoms, and other forms of local leadership. By using exiting governing structures, community members are more likely to respect and adopt the changes that are being proposed (CGIAR, 2013).

LITERATURE REVIEW

The essence of CBMRM is behavior change – getting people in the community to change their behavior to better conserve marine resources they rely on for their livelihood. There are essentially four basic ways to change behavior associated with natural resource conservation (Tyson, 2013). These include Dialogic strategies, Education strategies, Persuasive strategies, and Reinforcement strategies.

Focht (2005) prescribes that when the scientific community and the general public both have low rates of agreement on what needs to be done, the best strategy to use is a dialogic strategy. A dialogic strategy is often where most CBMRM projects begin. It involves bringing key people in the community that are affected by the problem into a discussion and having them work together to formulate an overall plan for what needs to be done.

Focht (2005) says that when the scientific community has a high rate of agreement on what needs to be done, but the general public does not agree much, the best strategy to use is either an Education strategy or Persuasion strategy. This includes running educational programs to increase knowledge about marine conservation issues if basic knowledge about these issues is not sufficient to begin with; or conducting persuasive communication campaigns to change attitudes and behaviors if adequate knowledge about the issues already exists.

Focht (2005) says that when the scientific community and the general public both have high rates of agreement on what needs to change, the best strategies to use are reinforcement strategies. This is often the final objective of a CBMRM project. This strategy includes community rules and laws that are enforced by the local government and/or incentives applied by market pressures.

In essence, the most freedom of choice for a community member is realized through dialogic strategies, followed by education strategies. Persuasion strategies give community members a bit less freedom to choose because at this point specific solutions have been decided upon and are now being promoted. Reinforcement strategies (laws and policies), give community members the least freedom of choice.

This paper will discuss the four strategies described above for changing behaviors and provide some examples of CBMRM activities in Fiji.

DIALOGIC STRATEGIES

Dialogic strategies are best used when there is a lack of consensus in both the scientific community and the general public (Focht, 1995). Dialogic strategies find solutions for situations when no solution is obvious. This is done by organizing community meetings with all community members that are affected or will be affected by the issue and having them discuss and compromise on possible solutions until one is decided on. Factors like short term and long term financial security and the ambiguity of hard science supporting remedial measures yields uncertainty that is best discussed openly. This process can take a very long time. Because of this, dialogic strategies are mainly used when long term sustainable solutions are what is needed.

Besides being labor intensive and time consuming, there are other problems with dialogic strategies. For example, public officials and experts may not see their role as non-biased educators but rather as persuaders trying to engineer the decisions that they want. Another barrier is when community participation in the decision making process is included late in the game to the point that it does not really have an effect on the outcome. In addition, lack of effective forums to bring all stakeholder together in a way that successfully educates these community members about the issues before any decisions takes place weakens the process (Cox, 2006).

Another term for dialogic strategies is collaborative learning. Collaborative learning emphasizes community education and conflict management. It is all about increasing community involvement in the decision process. Collaborative learning is thought to be different than some community participatory processes in that it tries to make it so that there is less competition between community groups. Collaborative learning promotes mutual learning and fact finding in the community. It allows underlying value differences to be explored. It uses fair and equal negotiation. It focuses on bettering the community rather than improving an individual's position. It allocates the responsibility for implementation across the entire community. With collaborative learning, solutions are found through an interactive and reflective process. It is often an ongoing process throughout the community. It is thought to handle individual and community conflict management, leadership, decision-making, and communication better than other strategies (Walker, Senecah and Daniels, 2006).

EDUCATION STRATEGIES

Education strategies are best used when the scientific community agrees with what needs to be done, but the general public is uncertain (Focht, 1995). Education strategies are used to give community members the knowledge needed to make well-educated decisions regarding their behavior (Archi, Mann and Smith, 1993). This is done by raising awareness of a situation, increasing knowledge about the situation, and teaching new skills. An important note about educational strategies is that they are not trying to change anyone's behaviors directly. The goal is to teach the target audience how to make informed well-thought out decisions, not tell folks what decisions they should make (Monroe, Day, and Grieser, 2000). Educational strategies are delivered in two ways, through formal education and non-formal education.

Formal education is generally done in a classroom setting. Audiences can be wide ranging in age, from youth to the elders of a community. The goals of formal education are to raise awareness of certain issues within the environment, as well as to give every person in the community the chance to obtain the knowledge and skills needed to deal with the environmental issue they face (Monroe, Andrews and Biedenweg, 2007).

Formal education can be broken down into four parts. The first is using presentations, books, videos, and field trips to provide the audience with information needed to make well-informed decisions about their behavior regarding the environmental issue. The second part is to increase the audiences' understanding of the environmental issue. This is done through group discussions, games, simulations, case studies, and experiments. This gets the audience involved in the learning process. The third part is about teaching new environmentally friendly skills or improving old skills in an environmentally friendly way. This is done through group skill building activities like community service projects and by answering any questions that anyone might not understand about the skills. The fourth step is helping those in the audience choose whether or not to act upon the information and skills provided to them. They must decide for themselves, but the instructor can still be there to answer any remaining questions and give advice if needed.

Non-Formal education is often used in conjunction with formal education to reinforce what formal education is teaching. Non-formal education teaches audiences how to improve skills to increase productivity in an environmentally friendly way and to help local governments develop sustainable economies and communities. It focuses more on hands-on engaging education. It often does not have to deal with the amount of bureaucracy that can be found in the formal education setting.

Non-Formal education generally takes place outside a classroom. One method is the use of extension education by programs based out of universities or government departments. An historical example of this in the U.S. are the agricultural research and educational programs run by land grant universities in every state. These programs are used to educate farmers on better resource management skills, pesticide control, soil testing, production practices and marketing skills (Association of Public and Land-grant Universities, 2011). The marine partner to this is the Sea Grant program run by the National Oceanic and Atmospheric Administration (Sea Grant, 2015).

A second method is the use of educational activities run by parks, zoos, aquariums, museums, and nature centers to give hands-on education (Kleis, 1973). Non-formal education can complement formal education efforts in the classroom or they can stand alone and not have any connection to formal education. Non-formal educators work with people of all ages and unlike with formal education, the audience is there by choice.

An example of this type of non-formal education in the U.S, can be seen at Mystic Aquarium in Mystic, Connecticut. Mystic aquarium has many programs that inform audiences about the effects of climate change and pollution on marine resources. These programs include guided tours, educational workshops, after school programs, summer camps, coastal clean-up days, population studies, and animal tagging. Mystic Aquarium also runs non-formal education programs that compliment formal education programs. They train teachers and supply materials for bringing marine education into the classroom. This is all done in an effort "to foster an interest in and understanding of the science and conservation of the aquatic world through formal and informal programs, written interpretive materials and exposure to living marine animals" (http://www.mysticaquarium.org/about/243-our-mission).

PERSUASION/SOCIAL MARKETING STRATEGIES

Persuasion/Social Marketing strategies are best used when there is a high amount of consensus among the scientific community, but a lack of consensus among the general public about that same issue (Focht, 1995). These strategies promote environmentally friendly behaviors in the same way that marketers influence people to purchase their products, hence the name Social Marketing. This strategy uses mass communication channels like social media, radio, and television to reach larger audiences as well as group meetings and face-to-face contact to reach smaller audiences. This strategy uses these communication channels to raise awareness and knowledge of the issue. Once the educational portion of the process has been established, efforts are switched from educating to changing the target groups' behaviors (Tyson, 2003).

Effective Social Marketing can be broken down into the following four-step process:

1. Campaign planning
   - Define issues and campaign objectives
   - Define audiences
   - Define messages
   - Define communication channels
   - Define information sources and strategic partners

2. Campaign implementation/management
   - Educating
   - Persuasion/Social marketing

3. Monitoring of campaign processes
4. Final outcome evaluation

Campaign Planning

The first step in planning a social marketing campaign is to identify the issue that needs to be fixed and the ways to solve it. The first stage in doing this is to learn about the issue, what it is, the causes of it, who it affects, etc. The second stage is identifying the path to achieving the final solution. This objective can be to raise awareness and knowledge of the issue or to create behavior change in the target audience.

The second step in planning a social marketing campaign is to identify target audiences. Target audiences can be grouped by age and occupation ranging from youth in the community, to businessmen, fishermen, politicians, or teachers. They can also be grouped by their geographic location, socio-economic status, education level, and media use, as well as similar levels of awareness, knowledge, interest, attitudes and behaviors on an issue. The best target audience for social marketing is found with the "least group size." The least group size is the largest amount of people needed to make a significant impact on the issue while still keeping it small enough to keep the campaign well defined and effective (Thelen, 1949).

The rate of a person's adoption of a new behavior is affected by their personal characteristics. Rogers (1995) grouped community members into five groups based on the rate that they adopt a promoted behavior. The fastest group of adopters are called the Innovators. This group is made of a very small portion of the population, between three to five percent. They can be quite eccentric and are not normally concerned with the social norms of the community. Though they do have the means and the resources to be able to adopt the new behaviors quickly, they do not have a large impact on others because they are not seen as well-grounded. The second fastest group of adopters are the Early Adopters. This group is made up of about 15% of the total population. Early adopters are normally well educated, have solid economic stability and are generally well respected by their peers. They have the resources to adopt early and because they are so well respected, become the most influential group in getting others to adopt as well. This makes them the best group for the campaigns to target initially. Once they adopt the promoted behavior, others tend to copy them.

The third group is called the Early Majority. This group is made up of about 33% of the population. They have less resources and ability to adopt new behaviors and thus are more cautious than the Early Adopters. They tend to observe the results of the Early Adopters before making any commitment. The fourth group is called the Late Majority. They represent another 33% of the population. Like the Early Majority, this group is cautious about adopting new behavior changes. They tend to change only when the rest of society already has. The final group is called the Laggards. This group makes up the final 15% of the population. This group is the hardest to convince to change. They are often older, poorer, or less educated than the previous groups and therefore are not capable of making a change easily.

The third step in planning a social marketing campaign is to create campaign messages that are aimed at the target audience. The challenge is to identify the benefits and costs of the behavior change that are most important to the target audience and develop campaign messages based on that. When creating campaign messages it is important to remember who the target audience is, the target audience's role in society, how their role affects the environmental issue, how committed they are to changing their behavior, the incentives that are being provided and the prompts that are in place to remind them of the message (McKenzie-Mohr & Smith, 1999). How committed the target audience is to the behavior change is extremely important. A person who has publicly committed to the cause and actively strives to change is much more likely to sustain the behavior change (Cialdini, 2001). Because of this, it is often useful to get the target audience to publicly announce their approval of the campaign message when the campaign is launched. This can be done by asking the target audience to participate in surveys and data collection about the campaign, asking the target audience to wear a button that shows support of the campaign, or asking the target audience to sign a pledge publicly supporting the campaign (McKenzie-Mohr and Smith, 1999). McKenzie-Mohr and Smith (1999) also recommend that the use of simple visual prompts be used as reminders of the campaign message. As an example, simple reminders like recycling bins with recycling logos on them that are placed in schools, offices, and public areas have been known to increase recycling in the community.

The fourth step in planning a social marketing campaign is figuring out which communication channels to use. These channels are divided into three categories, mass communication channels, group channels, and interpersonal channels. Mass communication channels include television, radio, internet, social media and handing out pamphlets. These channels are best used for creating interest, raising awareness or increasing knowledge of an issue with a large number of people. Group channels include field trips, community field days, and demonstration tours. These channels are best used for persuading the target audience to adopt these behaviors. Lastly there are interpersonal channels which mainly include face-to-face contact or over the phone contact. These channels are best used for reinforcing the adopted behavior. Each channel has its own pros and cons. For example television channels can reach anyone with access to a television and are very effective for creating interest and raising awareness through its content. However, television commercials are very expensive, have a limited amount of airtime, and its very hard to focus the message on a singular target audience. Group interaction such as at field days are very good at getting the audience talking with other community members so they can share their experiences and are good for teaching issue related skills. However, they lack the size of the audience that mass communication can reach and require a large time commitment from the target audience (Tyson, 2013).

Finally the fifth step in planning a social marketing campaign is finding strategic partners to help with the campaign. These partners can play important roles in the campaign by helping with funding, spreading the message, and providing expertise and help persuading the target audience. Groups that have been known to become strategic partners with social marketing campaigns include commercial sector distributors, commercial sector research agencies, commercial sector advertising agencies, governmental agencies, media (e.g., news stations, newspapers, and magazines), retailers, volunteer groups, funding agencies, other nonprofit organizations/advocacy groups, and large and small corporations. When recruiting partners it is best to emphasize the positive rewards of working with them and play down any costs of the partnership.

Tailoring Messages to Your Target Audience

In order for the campaign message to be properly developed, campaign managers need to know their target audiences' views on a few key variables. These are: Which resources are perceived to be the most threatened and which of these are most important to the community? What are the perceived positive and negative consequences of the proposed behavior change? What are the existing norms in the community and who is most likely to have an effect on these norms? How confident is the audience that the behaviors being promoted will be adopted, why/why not? Knowing the answers to these questions will allow social marketers to create and implement high quality campaign messages. The more the message is tailored to the audience, the more likely it is to be adopted. Other factors that have an effect on the likelihood of adoption of campaign messages include: message discrepancy and how credible the source of information seems to the target audience.

Message discrepancy is the difference between the target audiences' original position on the issue and the campaign's message on the issue. The greater the discrepancy, the lower the chances of the target audience adopting the desired behavior. This emphasizes how important it is that research be done to understand the target audience before the campaign is designed. If the challenge is to move the audience great psychological distance, the campaign may be best staged in a way that moves the audience in small steps toward the goal.

Source credibility is a combination of the perceived expertise of the campaign organizers and their perceived trustworthiness (Bettinghaus & Cody, 1994). Depending on the type of issue, different types of credibility are needed. Complex objective issues need a source with high levels of expertise. With risky or subjective issues, the audience needs high levels of trust in the people running the campaign.

Improving the Odds of Adopting Behaviors

Everett Rogers (1995) describes five factors that affect how well the desired behavior change is received. The first is relative advantage - this means that the new behavior must be seen as more profitable when compared to the economic, social, physical and psychological costs of making such a change. The second is compatibility - this means that the more compatible the new behavior is with traditional beliefs, attitudes, behavioral practices, and resources, the higher the chance of it being adopted. The third is complexity - this means that the simpler and more basic the new behavior seems to the audience, the easier it is for them to understand it, and this results in a higher chance of it being adopted. The fourth is trial-ability - this means that the more the new behavior can be tested out on a small scale before full adoption, the higher the chance of it being adopted. The fifth is observability - this means that the more transparent and concrete the final results of the behavior change are, the more likely the speed of adoption throughout the community will increase.

REINFORCEMENT STRATEGIES

As previously stated by Focht (1995), when scientists and community members agree on what needs to be done and have the proper knowledge skills, attitudes and intentions to behave correctly, laws and policies can be put into place to ensure that everyone acts according to plan. Reinforcement strategies are used to either encourage people to act correctly or to discourage them from acting incorrectly.

Positive reinforcement strategies can come in the form of rewards. For example, in the U.S. when state governments started implementing bottle and can deposits (five cents per bottle or can), they saw 68% less litter in Oregon, 76% less in Vermont and 82% less in Michigan (Mckenzie-Mohr and

Smith, 1999). By rewarding those who recycled, the goal of reducing littering was achieved.

Negative reinforcement strategies can come in the form of establishing a monetary penalty for those who do not behave correctly. For example, in Worchester Massachusetts a law was introduced that started charging people for every plastic grocery bag they used. This fee deterred people from using these non-environmentally friendly bags and reduced the waste stream by 45% (Mckenzie-Mohr and Smith, 1999). People started bringing their own reusable grocery bags.

Legal strategies are fairly straightforward. Governments create laws that regulate or punish those who do not cooperate with the desired behavior. These strategies are usually used for more serious matters, like when poaching is a big problem.

Encouraging and discouraging reinforcement strategies are often the strategies of choice when facing what is referred to as a "Social Dilemma". Social Dilemmas are common when trying to conserve common property that no one owns but everyone benefits from, like a fisheries. A Social Dilemma is the idea that each person will try to achieve the most amount of personal gain with the limited amount of time and money that they have; i.e., they will catch as many fish as they possibly can with the least amount of effort. With short foresight and bad management practices a few people can ruin a natural resource that many rely on just to maximize their own personal gain. The dilemma is that a person is personally better off for not cooperating with the conservation efforts of the community, but if many do not cooperate than the resource will crash and everyone will lose (Dawes, 1980). There are generally two reasons why a social dilemma can exist. First, there is the fear that if everyone does not change their behavior which seems doubtful, it would be a needless sacrifice for an individual to change. Secondly, there is the idea that if a large enough majority has already changed their behavior and is conserving the resource, than an individual can cheat without having any great effect on the resource (Weiner and Doescher, 1991). As stated previously, Social Dilemmas are normally solved with incentives to reward the desired behavior, penalties to discourage bad behaviors, and laws and regulations to control everyone's behavior.

COMMUNITY BASED MARINE RESOURCE MANAGEMENT (CBMRM) IN FIJI

In Fiji, the problems being faced include, overfishing, over collecting of shells, poaching, climate change, and habitat destruction. The main CBMRM strategies used to address these problems include dialogic, education strategies, persuasion/social marketing strategies and reinforcement strategies.

Education Strategies

Besides conducting transect surveys of marine conditions, volunteers on the GVI Caqalai Base interact with local communities on the neighboring island of Moturiki. Volunteers teach in the Uluibau Primary School and the Moturiki District School. They are also able to help out on villagers' farms on weekends, help out on local seaweed farms, and stay overnight in the villages for ceremonial kava sessions. This allows volunteers to interact and befriend members of the local community and gives them a chance to share what they have learned about the reef. In addition to volunteers going to the villages, the villagers also come to Caqalai to learn from the volunteers. These interactions were modeled around both the formal and non-formal education strategies discussed earlier.

<u>Formal Education</u>

Most formal education being done by GVI is performed in primary schools. This can be broken down into three parts. The first part is providing the children with information. Every class starts off by giving a small lecture about the topics the children would be learning that day and a small review of the topics they had been taught the previous week. These topics have included, learning about sharks, fish, invertebrates and other marine animals on the reef, learning about recycling, upcylcing, proper management of waste and how waste affects the oceans and marine life, the different types of ecosystems in Fiji, weather patterns, and good personal hygiene techniques. Once the initial presentation has been completed, understanding of the topic is reinforced through word games, like hangman, crosswords and matching games, art projects, that include posters and pictures they could hang in their classroom or bring home, and outside activates, like beach clean-ups, field trips and running games. Students are encouraged to ask questions at any time throughout the class regarding any of the subjects they have learned. At the end of each trimester the students review the topics they have learned over the past few months and take a final exam. The exam helps to reinforce the topics they learned as well as assess how well the students learned the topics. Every student taught by GVI at Uluibau Primary School and Moturiki District School passed the exams for every subject they were taught.

Field trips and beach clean-ups are especially effective when teaching about waste management, plastics in the oceans, and other marine related topics because they give the students real life examples of the topic they are learning about. This is important when working with the younger students who might not fully understand the lessons because of the language and vocabulary. Beach clean-ups also increase how much students care for their environment. Seeing firsthand the garbage along the beach and being given an explanation about the effects this garbage has on the reef ecosystem clearly depicts the importance of properly disposing of their garbage.

While most of the older students in grades 5-8 are able to understand the majority of the information they are taught in the presentations, the younger students in grades 1-4 sometimes have difficulty. This is mainly because of the language barrier. That being said, the younger kids are still able to learn a lot of what is taught because of the practical exercises. Fijian schools emphasize the use of English language during class time and this makes it difficult to teach younger children who have a weaker command of the language. In addition, the more complicated subjects include lessons on the science behind the water cycle, the role of phytoplankton in the oceans, and ocean acidification due to climate change. Lack of understanding of science and the scientific terminology associated with these issues, coupled with the fact that the students cannot physically see theses issue, makes these subjects difficult to teach. Subjects like the role of algae grazing fish and top level predators like sharks on the reef ecosystem and the damages caused by plastics and human waste in the oceans are more easily understood by the students because they have seen these things first hand and have experience dealing with them.

An added benefit of education in the primary schools is that the children bring the information back to their parents. The parents of these children are fishermen, shop owners and villagers that rely on the marine environment for their livelihood. Learning from their children peaks their interests in knowing more about the reef and how to protect it.

Non-Formal Education/Dialogic Strategies

Besides formal education in the local primary schools, GVI also holds non-formal education presentations for local adults. In the past, men (ages 18-30) and some of the elders were invited to the Caqalai base to learn about water quality, plastics in the oceans, climate change, effects of over fishing, the importance of the different types of fish and corals and the importance of mangroves and sea grasses. Presentations are made using PowerPoint with pictures, diagrams and examples. These presentations are followed by a lengthy question and answer period with the scientific members of the GVI staff. A film documentary is shown about the effects of plastics in the oceans and a beach clean is conducted around Caqalai. The participants are also given the option to go on a snorkel outing to see and learn about the reef.

Non-formal education is done by GVI in two other ways. The first is by augmenting school-based formal education with trips to the GVI Caqalai Base for primary students. The school trips are an excellent way for the children to see firsthand what the GVI volunteers do on the reef. The students play marine ecosystem related games, listen to presentations and stories about what GVI does on the reef, and go snorkeling. Despite living right next to the reef, this is the first time for many of the kids to see the reef underwater and it is an excellent way for them to experience the beauty of it.

There are also trips by GVI volunteers to local villages where they address environmentally responsible ways to create income. These trips give the volunteers a chance to interact with the villagers and teach them about the importance of their reef and what they can do to help protect it. One example is helping to develop seaweed farms. Certain types of seaweed are farmed for chemicals found in them. These chemicals are used to make cosmetics, toothpaste, and almost every type of gelatin-based food. The demand for these seaweed chemicals is very high, especially in Asia. Farming seaweed is inexpensive and requires very few materials (e.g., seaweed saplings, string, wooden stakes, and an open plot of water preferably with a sandy bottom). All of the materials except for the plot of water are subsidized by the Fijian government in an attempt to stimulate income generation in an environmentally responsible way. However, it is a very labor intensive enterprise. GVI volunteers travel to the local villages that want to set up these farms and help tie the saplings to the lines, plant the stakes in the water, tie the lines to the stakes and help maintain and clean the lines. Working alongside local community members provides an excellent opportunity to talk about the issues they have concerning the environment and answer any questions about the environment they might have.

Persuasion/Social Marketing Strategies

While GVI mainly focuses on education strategies with local communities, persuasion/social marketing strategies are also used by conservation programs in Fiji that GVI indirectly supports. The 4FJ program stands for "For Fiji" and is a program that aims to stop the consumption, sale, purchase or catching of grouper during their mating season. These fish are some of the most important food fish in Fiji and are a major source of protein for a large majority of people. Grouper population numbers have dropped dramatically in recent years, primarily due to overfishing during the mating season. Grouper congregate in the same area every year to reproduce in large schools, making them very predictable and easy to catch. Before the 4FJ movement the fishing season was centered on the months that these groupers schooled and spawned. Unfortunately and predictably, as the population of Fiji has increased so has demand for Grouper, and Grouper population numbers have dropped drastically.

The 4FJ program is a social marketing campaign that tries to persuade people to give the Grouper an opportunity to effectively reproduce so populations can recover from years of overfishing. The program tries to achieve this goal by getting people to make a public pledge stating they would not consume, sell, purchase or catch Grouper during their mating season. The target audience of this program is essentially every person in Fiji because almost everyone consumes Grouper caught by fishermen who depend on it for their livelihood. The 4FJ program uses mass communication channels such as TV, radio, billboards, and social media (public online pledges) to reach out to the Fijian public. This program has worked well for villages on the Viti Levu and Vanua Levu, as they are the two biggest and most developed islands, but the smaller islands need more specific and personal communication channels. GVI volunteers help with this. Through face to face interactions at school presentations and ceremonial kava sessions, the 4FJ movement, with GVI assistance, has been able to gain the support of many of the local chiefs and local schools on the smaller islands.

On the main island, the 4FJ program has been able to gain support from many important Fijians that people look up to and respect. The two biggest "idols" the 4FJ program was able to recruit to help are the Fijian president, George Konrote, and the King of 7s Rugby, Waisale Serevi. Rugby is a hugely popular sport in the Fijian culture as almost everyone grows up playing it or watching it. Gaining the King of 7s Rugby's support was a real win for the 4FJ project. Other notable celebrities that 4FJ was able to secure support from include Miss Fiji and various Fijian television and radio hosts. Those who pledged to not consume, sell, purchase or catch grouper during their mating season were given either a badge, sticker, business plaque, sign or a facebook badge to publicly show that they took the pledge. This public pledge not only reinforces the pledgers' likelihood that they will act on their pledge, but also helps convince those around them to take the pledge through normative pressure. Much of the Fijian population has publicly taken the pledge and those that have not, feel peer pressure to do so. It is fully expected that down the road, the vast majority of the population will eventually pledge to the 4FJ program.

Reinforcement Strategies

Brandon Paige, a South African shark expert and behaviorist, founded the Aqua Trek Shark Dive in Pacific Harbor in 1999. This is an operation that GVI uses for its interns to complete a three month attachment and earn their Master Diver certification. The shark dive is one of the only places in the world where a diver can see seven different species of shark on one dive. The species of sharks that can be seen include bull sharks, tiger sharks, lemon sharks, silver tip sharks, white tip reef sharks, black tip reef sharks and tawny nurse sharks. Brandon's goal when founding this shark dive was to educate divers to the beauty of sharks and to show that they are not the cruel killing machines that the media often portrays them to be. It is one of the only places in the world that a diver can see wild sharks being hand fed, up close, with no cage.

The dive operation takes place on protected fishing grounds. Fiji has always had a tradition of locally protected fishing zones. Coastal villages have always had the sole fishing rights to the reef just off their coastline and have been able to choose who they permit to fish these areas. The government has officially preserved these traditional rights. When Aqua Trek arrived in 1999 the reef they chose to develop the shark encounter operation on was completely dead because of years of overfishing. For many years, long liners and other commercial fishermen had been sold licenses to fish the reef with no limit. This resulted in the depletion or death of algae grazing fish, sharks, commercial fish, invertebrates and corals. Aqua Trek struck a deal with the villages that owned the traditional fishing rights to the area that if they stopped the sale of fishing licenses and created a protected marine park, Aqua Trek and other dive companies would pay them a marine park fee for every diver they took on the reef. The villagers were initially skeptical but agreed to a trial period to see if it might be commercially viable. In the first month that Aqua Trek operated the shark dive, the village made significantly more money than they had selling fishing licenses.

Aqua Trek's ecotourism has helped convince the government to declare most of the reef in the Beqa Passage and Beqa Lagoon a national marine park area and has stopped long lining and overfishing in those parks. The villages have been able to use the money collected from ecotourism to repair roads, improve and repair schools, and repair other infrastructure in the area. Seeing this success, neighboring villages have also adopted this method of protecting their reefs and have worked to develop similar ecotourism businesses.

Aqua Trek also struck a deal with the local fish factory in Fiji's capital, Suva. Normally the fish factory in Suva had no use for scrap tuna heads, skins and loins and would normally discard them. Instead of wasting these parts of the fish, Aqua Trek recycles them back into the ecosystem. The fish heads are hand fed to the sharks and the skins and loins are used to feed the abundance of fish species that congregate around the shark dive. This provides a food source for fish that otherwise, because of the poor health of the coral on the reef, would be undernourished. Through the feeding of sharks and fish species, the reef ecosystem in the area is recovering at a very quick rate. With healthy fish (including sharks), the food cycle on the reef has started to improve. When fish are not feeding regularly they are only able to reproduce about once a year. When fish populations are feeding regularly they are able to reproduce three to four times a year. The recovery of the fish populations has trickled down and the coral populations are visibly healthier than they were when Aqua Trek first started the dive operation. In addition, local fishermen have reported catching more fish and larger fish outside the marine park area since the shark dive started.

In addition, associated with the deal to purchase the discarded parts of the fish, Aqua Trek has been able to convince the fish factory to stop fishermen from long lining near the marine park area. The marine park area only protects a certain amount of area, but long liners and commercial fishermen are still able to fish just outside of the park. Most of the bigger sharks are not resident sharks in the park (i.e., tiger sharks, bull sharks, and lemon sharks). They are known to travel hundreds of miles up and down the coast to hunt and travel even greater distances to their traditional breeding and birthing grounds.

The shark dive operation is a perfect example of CBMRM. It optimizes the relative advantage of cooperating with the program, as it is ultimately more profitable to the community than allowing commercial fishermen to overfish the area. It addresses compatibility and complexity as Aqua Trek works with the villages to use their traditional fishing rights to protect their reef and the concept is easy to understand. The results are easily observable as divers willing to pay money to scuba dive converge on the area. Aqua Trek also factored in the attribute of trial-ability by first allowing villagers to see in the short-term how they can make more money protecting the sharks and the reef from commercial fishing. Once, the operation was a proven success, the villagers accepted this practice permanently. They have now been protecting the sharks and the reef like this for the past 16 years.

BIBLIOGRAPHY/REFERENCES

The Situation in Fiji

Global Vision International (2011). *Marine Conservation Ecology* in Fiji Marine Conservation Expedition Training Manual.

Research Methods for Observing Marine Species

Beruman, Michael L., & Barun, Carmin D. (2014). *Movement Patterns of Juvenile Whale Sharks Tagged at an Aggregation Site in the Red Sea*. PLOS ONE.

Caratti, John F. (2006). *FIREMON - Fire Effects Monitoring and Inventory System*. Fort Collins, CO: U.S. Dept. of Agriculture, Forest Service, Rocky Mountain Research Station.

Chin, A., Lison De Loma, T., Reytar, K., Planes, S., Gerhardt, K., Clua, E., & Burke, L., Wilkinson, C. (2011). *Status of Coral Reefs of the Pacific and Outlook*. Global Coral Reef Monitoring Network (GCRMN).

English, S., Wilkinson, C., & Baker, V. (1997). *Survey Manual for Tropical Marine Resources*, 2nd Edition. Australian Institute of Marine Science.

Global Vision International (2011). *Marine Conservation Ecology.* Fiji Marine Conservation Expedition Training Manual.

Harewood, Asanchia, & Julia Horrocks (2008). *Impacts of Coastal Development on Hawksbill Hatchling Survival and Swimming Success during the Initial Offshore Migration.* Biological Conservation 141.2, 394-401.

Hill, Jos, & Wilkinson, Clive R. (2004). *Methods for Ecological Monitoring of Coral Reefs: A Resource for Managers.* Australian Institute of Marine Science.

Labrosse, Pierre, Michel Kulbicki, & Jocelyne Ferraris (2002). *Underwater Visual Fish Consensus Surveys.* Reef Resources Assessment Tools.

Wilkinson, Clive (2004). *Status of Coral Reefs of the World, Volume 1.* Australian Institute of Marine Science.

Marine Species of Fiji

Allen, Gerald, & Roger Steene (2005). *Reef Fish Identification for the Tropical Pacific.* Jacksonville, United States: New World Publications.

Foy, Sally (1982). *The Grand Design: Form and Colour in Animals.* Oxford Scientific Films. BLA Publishing Limited for J. M.Dent & Sons Ltd, Aldine House, London.

Humann, Paul, & DeLoach, Ned (2010). *Reef Creature Identification for the Tropical Pacific.* Jacksonville, United States: New World Publications.

Fiji Department of Environment (2010). *Fiji's Fourth National Report to the United Nations Convention on Biological Diversity.*

Global Vision International (2011). *Marine Conservation Ecology.* Fiji Marine Conservation Expedition Training Manual.

Jokiel, P. L., & S. L. Coles (1990). *Response of Hawaiian and Other Indo-Pacific Reef Corals to Elevated Temperature.* Web retrieval, Dec. 2015.

MacTavish, Thomas, Stenton-Dozey, Jeanie, Vopel, Kay & Savage, Candida (2012). *Deposit-Feeding Sea Cucumbers Enhance Mineralization and Nutrient Cycling in Organically-Enriched Coastal Sediments.* PLOS ONE. Public Library of Science, 27 Nov. 2012.

Rice, J., Cooper, J., Medley, P. & Hough, A. (2006). *Surveillance Report South Georgia Patagonian Toothfish Longline Fishery.* Moody Marine.

Snyderman, Marty (2009). *Underwater Propulsion: A Tale of ot Tails.* Dive Training Magazine.

Tanden, E. M. (2008). *Pelvic Fin Locomotor Function in Fishes: Three-dimensional Kinematics in Rainbow Trout (Oncorhynchus Mykiss).* Journal of Experimental Biology 211.18: 2931-942.

Tyson, C.B. (2015). Personal n otes taken while surveying and studying marine species in Fiji from June to December, 2015.

Worm, B., et al (2006) *Impacts of biodiversity loss on ocean ecosystem services.* Science, 314: 787.

WWF. (2015). *Shark.* WorldWildlife.org.

Community Based Marine Conservation

Archie, M., Mann L., & Smith, W. (1993). *Environmental Social Marketing and Environmental Education.* Academy for Educational Development, Washington, D.C.

Association of Public and Land-grant Universities (2011). *Land-grant Heritage.* [ONLINE] Available at: http://www.aplu.org/page.aspx?pid=1565. [Last Accessed October 23, 2012].]

Bettinghaus, Erwin P. & Cody, Michael J. (1994). *Persuasive Communication* (5th Edition). Fort Worth: Harcourt Brace College Publishers.

Cialdini, Robert B. Influence: Science and Practice. Boston, MA: Allyn and Bacon, 2001.

Cox, R. (2006). *Environmental Communication and the Public Sphere.* Thousand Oaks: Sage Publications.

Dawes, R. (1980). Social dilemmas. *Annual Review of Psychology,* 31, 69-93.

Focht, W. (1995). *A Proposed Model of Environmental Communication Ethics.* National Association of Professional Environmental Communicators Quarterly, Spring issue, 8-9.

Kleis, J., Lang, L., Mietus, J.R. & Tiapula, F.T.S. (1973). *Toward a Contextual Definition of Non formal Education.* Non-formal education discussion papers, East Lansing, MI: Michigan State University.

McKenzie-Mohr, D. & Smith, W. (1999). *Fostering Sustainable Behavior.* New Society Publishers, British Columbia.

Monroe, M., Andrews, E., & Biedenweg, K. (2007). *A Framework for Environmental Education Strategies.* Applied Environmental Education & Communication, *1*, 13. Retrieved July 18, 2012, from http://envacapstone.wiki.usfca.edu/file/v.

Monroe, M., Day, B., & Grieser, M. (2000). *Environmental Education & Communication for a Sustainable World Handbook for International Practioners.* Washington, DC: Academy for Education Development.

Mystic Aquarium (2015). *Mystic Aquarium - Our Mission.*

National Oceanic and Atmospheric Administration (2015). *Sea Grant.* Available at: http://seagrant.noaa.gov/.

Ostrom, E. (1990). *Governing the commons - the evolution of institutions for collective action.* Cambridge University Press: New York.

Rogers, E. M. (1995). *Diffusion of Innovations* (fourth edition). New York: The Free Press.

Tyson, B. (2013). *Social Influence Strategies for Environmental Behavior Change.* IUniverse Publishers, Bloomington, IN.

Walker, G., S. Senecah, and S. Daniels. 2006. *From the Forest to the River: Citizens' Views of Stakeholder Engagement,* Human Ecology Review 13(2):193-202.

Wiener, J. L. & Doescher, T. A. (1991). *A Framework for Promoting Cooperation.* Journal of Marketing, 55, 38-47.

Worldfish (2013). *Community based marine resource management in the Solomon Islands.* CIGAR Research Program on Aquatic Agricultural Systems. Penang, Malaysia. Manual-2013-17

INDEX